Risk Monetization

**Converting Threats and Opportunities
into Impact on Project Value**

Risk Monetization

Converting Threats and Opportunities into Impact on Project Value

Glenn R. Koller

CRC Press
Taylor & Francis Group
Boca Raton   London   New York

CRC Press is an imprint of the
Taylor & Francis Group, an **informa** business

A CHAPMAN & HALL BOOK

CRC Press
Taylor & Francis Group
6000 Broken Sound Parkway NW, Suite 300
Boca Raton, FL 33487-2742

© 2012 by Taylor & Francis Group, LLC
CRC Press is an imprint of Taylor & Francis Group, an Informa business

No claim to original U.S. Government works

Printed in the United States of America on acid-free paper
Version Date: 2011912

International Standard Book Number: 978-1-4398-1829-9 (Hardback)

Library of Congress Cataloging-in-Publication Data

Koller, Glenn R. (Glenn Robert), 1951-
  Risk monetization : converting threats and opportunities into impact on project value
/ Glenn R. Koller.
        p. cm.
  Includes bibliographical references and index.
  ISBN 978-1-4398-1829-9 (alk. paper)
  1. Risk management. 2. Uncertainty. 3. Money. I. Title.

HD61.K634 2012
658.15'5--dc23   2011023788

Visit the Taylor & Francis Web site at
http://www.taylorandfrancis.com

and the CRC Press Web site at
http://www.crcpress.com

Contents

Preface

Why should anyone read this book? One word—"money"! Regardless of how we generate our paychecks—as an entrepreneur or as part of a nonprofit, governmental, educational, or corporate entity—we all strive to make and save money. This book offers practical guidance with regard to how to identify, assess, manage, and monetize risks—both threats and opportunities—so that risk will have as positive an impact on the "bottom line" as is practical and possible.

In one of my previous books—*Modern Corporate Risk Management: A Blueprint for Positive Change and Effectiveness* (see Reference section at the end of this preface)—I begin the Introduction with the heading: It is All about Changing Behaviors. This is true enough, and in that book, I delineate in detail the issues and challenges related to behavioral change within the political, cultural, and organizational context of any corporation, academic institution, or governmental agency.

Given the significant barriers to behavioral change and considering the arduous and painstaking methods and techniques—described in the aforementioned book—that need to be deftly applied to realize positive behavior modification, one might ask oneself: "To what end?" Although it is true enough that prompting an organization to accept and implement a cogent risk process is centered on behavior modification, making and saving money are the primary motivations behind implementation of a risk assessment, management, and monetization process.

This book is not a novel. Therefore, I expect few readers to start reading on the first page and continue reading to the last. I am well aware that books such as this one are typically utilized as a reference material. That is, a person wants to know something about, say, "chance of failure" and so they turn to the book's index, look up the subject, and read only the indicated pages. Being cognizant of this behavior has led me to emphasize certain critical themes and subjects numerous times in the book—this is by design.

Just for example, the concept I refer to as "One, Two, Three—Go!" is described or eluded to in several sections of the book. I return to this theme, and others, for at least two good reasons.

First, it is important. Knowing that readers will use the book mainly as reference material, I wish to ensure—as best I can—that regardless of what major theme is researched through an index search, the reader will at least see reference to important topics. If such concepts and techniques were described or eluded to only once, such an important aspect could easily be missed.

Second, repetition of an important concept in various contextual settings is a time-honored method for causing the concept to "stick" in the mind of the reader. Returning to the same concept in different guises is one of the most effective means of assuring that the reader will retain the premise.

Again, while it really is all about changing human behavior, the reason for attempting to change the behavior of employees and management is to create a more efficient and, therefore, more profitable enterprise. Faithfully following the steps described in this book that lead up to risk monetization, and the process of monetization itself, will help ensure that projects—regardless of their ilk or the type of organization to which they are germane—will be as trouble free and as profitable as is practical to expect. Relating to the reader how to convert probable pertinent threats and opportunities, that is, risks, into positive impact on the chance of success and/or profitability of any project is what this book is all about. If you draw a paycheck, then you should consider reading this book (maybe first page to last!) and attempt to implement the precepts in your organization.

In the Appendix of this book, I have attempted to summarize the most significant aspects of the processes leading up to risk monetization and the monetization process itself. This summary in no way is meant to substitute for the critical details embodied in the main text, but is a great way to introduce the subject to those who might need to be convinced that implementation of a risk-monetization process is necessary in your organization. Although a bit more than an "executive summary" and significantly less than the entire text, the spoon-feeding of the Appendix sections—perhaps one at a time over a short period—can help introduce the critical concepts of risk monetization and can aid in convincing any organization that implementation of such a process will be immensely beneficial. Good luck, and feel free to contact me directly if you have more questions on the subject.

Acknowledgments

I thank Ian Bryant for granting permission to publish this book.

I thank David Pells, Managing Editor of PM World Today, for his permission to reproduce material from PMWT articles in the appendix of this book.

I am, as always, ever grateful to Bob Stern, Publisher at Taylor & Francis, for years ago taking a chance on me and my first book. I will never forget that.

As with every book, I am eternally grateful to my wife Karen for her indulgence while I write these books. Last one, hun, I promise!

Introduction

As with all of my other books, this treatise is written in layman's language. That is, I assume that the user is not necessarily an expert in the ways of risk assessment, management, or monetization. Any material that is not covered in this book, such as the basic concepts behind Monte Carlo analysis, is referenced in other books I have written on the subject.

In addition, this book is not aimed at any particular type of enterprise or area of expertise. I attempt to write these books such that anyone serving in nearly any capacity in a nonprofit, government, educational, or corporate setting will be able to comprehend and practically implement the concepts and techniques conveyed.

Why the Book Is Organized as It Is

It always is my aim to write a book that can be picked up and effectively read by anyone, to convey to that person the premises that pertain to the subject matter, and to instill in the reader the motivation and confidence required to practically help them implement the necessary processes and techniques.

To that end, I have organized this book in what I consider to be a logical progression of steps and concepts. That is, the progression of subjects appears logical to me.

To the reader, it might seem odd that I have put off until Chapter 3 definitions of the fundamental terms and concepts. Odd though it might seem, there is an underlying method to my madness.

Converting risks—probable and pertinent threats and opportunities—into impact on project chance-of-success and/or value is not something that is done consistently or routinely. The breaking of routine necessarily means introduction and implementation of new and, sometimes, uncomfortable concepts and methods. If in the first chapters I would have attempted to convey those seemingly unorthodox precepts and techniques without setting the stage for their necessity, I presume most readers might reject the necessary deviations from the norm and read no more of the book.

An old adage says, "You can't make an omelet without breaking a few eggs." The omelet I will attempt to make in this book is the description of how a risk-monetization process can be exceedingly beneficial to any project in nearly any context. The eggs to be broken are some of the contemporary workflows in organizations and many of the commonly held beliefs about and definitions of risk-related concepts and terms.

Details of the Book's Organization

As a hedge against reading the first part of the book and finding it so different from the contemporary precepts that it is rejected out of hand, I have implemented at least two things. First, I have attempted in Chapters 1 and 2 to relate why a risk-monetization process might be highly beneficial and to describe how risk matters typically are less-than-adequately handled in contemporary organizations.

"What's the Point?" is the title of Chapter 1. In the major sections (and many subsections), namely,

- A Focus on Success and Value—*Not* on Reducing Uncertainty
- Many Contributing Disciplines
- Many Views of Risk
- An Aim to Integrate Risks
- Surprise! Integrated Impact Is Almost Always Alarming
- Where We Would Like to Be

I make the case for why annealing of myriad views of risk is necessary and practical. I also illustrate why bring to the fore the often surprising fact that it is only through the integration of risks from all disparate disciplines that we get the most realistic view of the impact of risks on project success and value. I also, in the end, describe the utopian risk state for which we should all be striving and what it might take to achieve that state.

In Chapter 2 entitled "How Risk Information Traditionally Is Handled—The Good, the Bad, and the Ugly," I attempt to describe the contemporary usage of risk registers. Although fine and well-intentioned instruments, I describe each of the primary elements of a fundamental risk register and suggest how each aspect might better be described and handled. Also delineated are the additional elements of a risk register needed to implement a risk-monetization process.

My aim in the first two chapters is to convince the reader that the step of risk monetization—the conversion of probable pertinent threats and opportunities into impact on project success and/or value—is a critical element of successful project execution. Having hopefully accomplished this task, it is in Chapter 3 that I present the terms and definitions that will be required to be accepted if risk monetization is to be a reality in a project.

It is not that the terms and definitions in Chapter 3 are completely "off the wall" with respect to conventional thinking about risk, but it can be a leap in an uncomfortable direction, for example, to accept that risks can be represented by both threats and opportunities. Thus, the setting of the stage in Chapters 1 and 2.

It is typical that readers believe that all of this risk stuff is great for other people, but it could not apply to their particular situation. The subject of Chapter 4—"Spectrum of Application for Monetization"—is to attempt to convey to the user that risk-monetization techniques can be applied in nearly every project setting. Details of several project types are offered as examples.

If readers cannot "see themselves" in the processes described—that is, if the situations delineated don't "speak to them" and the reader can't relate to the subject—then most elements of a book like this one will seem to the reader to be abstract and impractical. In Chapter 5—"Without Risk Monetization, What Typically Is Done"—I convey the quintessential risk situation for most projects. It is my aim that while digesting this chapter, each reader will wonder how I got inside their organization because the descriptions given quite accurately match how things are done in their area. The shortcomings of these conventionally utilized processes are emphasized.

Well, if we could start from scratch, so to speak, and build a risk process in a "virgin" environment, just how would one go about doing that? That's the subject of the rather lengthy Chapter 6. In this chapter, I describe in detail exactly how risk identification, assessment, management, and monetization processes might be set up in an environment in which the risk-process proponent is brought in at the inception of the project and is able to perfectly implement the various risk processes. This is the "high control" or "perfect world" scenario.

In Chapter 7, I relate how a risk-process proponent might handle the less-than-perfect-world situation. That is, it can be typical that a formal and comprehensive risk process is only called for when a project realizes it is in trouble. At such a stage, various disciplines might have already implemented disparate risk processes—that will need to be integrated—while other critical areas of expertise have done little or nothing about risk at all. Following the steps outlined in this chapter and in Chapter 6 will effectively set up the risk-monetization step regardless of the starting disposition of the project organization.

Finally, in Chapter 8, the risk-monetization process is described. An example is offered. Given that a primary motivation of most project teams is to make and save money, the justification of risk monetization in that context is well established. The rationale behind the concept that all risks should not be addressed and rationales behind other critical elements are completely and practically described. How investment decisions—based on the monetization and ranking of risks—are made is a primary focus. Adherence to the precepts outlined will result in more successful and profitable projects.

As mentioned in the preface of this book, the long process leading up to risk monetization and the monetization process itself is rife with critical details. However, such processes might never "get off the ground" if decision makers can't be convinced regarding their necessity. Shoving a book like this one under their noses is not likely to positively sway them. The abbreviated sections in the Appendix of this book are *not* designed to be a guide for implementation of a risk-monetization process. Rather, the short sections are meant to contain just enough, and just the right information, to compel readers to investigate and, hopefully, implement a risk-monetization process. Presenting the Appendix sections en-masse or as "weekly install-ments" might be just the documentation required to get the risk conversation started.

Author

Dr. Glenn Koller currently is a risk advisor at a major corporation. Glenn retired after 30 years from a major energy company in 2008 as the head of that company's Major Project Risk Group. He received his PhD in geochemistry/geophysics from Syracuse University. His current and past responsibilities include all aspects of risk and uncertainty analysis, management, monetization, and training. Responsibilities of his current position include implementation of risk-assessment and risk-management technologies and processes in the corporation; development of statistical routines that comprise corporate risk systems, keeping abreast of risk technology developed by vendors, other companies, and the national laboratories; marketing risk technology; and performing technical and consulting services. Glenn holds two international patents on statistical routines (three patents currently pending). This is the fifth book Glenn has authored on risk/uncertainty. He currently resides in Tulsa, Oklahoma. Glenn's e-mail address is riskaid@cox.net.

1

What's the Point?

Skipping the Fundamentals for Now

It might seem logical and smart—and I've never been accused of that—to at the beginning of this book define some of the critical terms that will be repeatedly utilized (the term "risk" for example). I have made a conscious decision to delay such definitions until Chapter 3 (no peeking, now!). So, you're thinking, "Why would this guy start writing about things that he is yet to define?" In my tiny and twisted mind, there is a compelling rationale to do so.

Most large organizations are composed of multiple disciplines. A corporation might contain separate entities that handle commercial, financial, legal, technical, engineering, security, environmental, logistical, health and safety (H&S), and other aspects of a business. If one were to approach a denizen of any of these corporate cul-de-sacs and ask that person to relate what risk means to them, one would end up with as many definitions of the term as people queried.

For example, it is not uncommon for personnel in the security area to see most risks as threats. These are things to be identified and eradicated. Representatives from the finance department, however, are likely to view risk as an opportunity—lower risk, lower reward and higher risk, higher return.

When approaching the subject of risk from an organization-wide perspective, the definitions of risk and of other terms have to be ones that can be universally embraced and employed. My definitions of risk and other critical terms have this universal applicability. However, most readers of this text will likely have backgrounds in one of the aforementioned disciplines (commercial, financial, legal, etc.) and, therefore, will harbor a view of risk that is typical of that arena—that is, colloquial.

If I were to hit the reader with my universal definitions of critical terms, the reaction might be to dismiss those descriptions because my characterizations might not square with the reader's discipline-specific viewpoint. Therefore, I will use the first two chapters of this book to gradually, and hopefully logically and convincingly, build a case for the critical-term definitions I would encourage the user to embrace.

In classes I teach, I always employ the analogy of teaching someone to doing a backstroke. If you are an experienced swimmer and have been doing the backstroke "your way" for years, then any attempt on my part to have you abandon your method and to embrace my technique likely will meet with resistance on your part. Some of that resistance will be natural in that your brain is wired to do it your way, and any change in "automatic" behavior is a difficult one. Push-back will also stem, however, from the fact that you believe that your backstroke method is more than adequate and you don't see the advantage in fixing what does not seem to be broken.

Another ubiquitous source of resistance to change stems from the fact that most people who deal with risk in their professional lives do so in the relative shelter of a specific discipline. That is, some folks in the insurance industry would swear that their use of Beta and other insurance-specific terms and concepts is all there is to understanding risk. Likewise, employees of environmental-analysis firms would attest to the fact that their attention to exposure time and method, pollutant potency and concentration, population densities, and the pathways by which a contaminant can spread would constitute a complete and holistic comprehension of risk. In neither of these cases—and in no other discipline-specific analyses—do the practitioners consider the truly holistic set of risks that can emanate from political, cultural, legal, organizational, logistical, commercial, financial, security, and so many other sources.

One of the main tenets of this book, and of all of my other books, is that we have only a relatively good grip on risk when we consider the entire spectrum of disciplines that can potentially impact a project. It is natural for any person in any specific discipline to view risk in a colloquial fashion. When a more broad-spectrum view is taken, however, it is almost always the case that the terms, definitions, and concepts that serve a particular discipline so well will fall woefully short of being potent enough to be used universally, that is, across disciplines. It is highly unlikely, for example, that the risk-related terms, definitions, and concepts employed by structural engineers would be embraced and employed by lawyers.

The trick I have to pull off here is twofold. First, I have to convince the somewhat discipline-biased reader that there is a case to be made for the integration of risks from all sources and that the terminology and methods employed from any single source, that is, discipline, likely will not suffice for the process of addressing an integrated set of multidisciplinary risks. Second, I have to attempt to make a credible case for the terms, definitions, and methods that can be universally applied to risks from any discipline or to an integrated set of risks from multiple disciplines. This is no minor undertaking!

So, whether you consciously realize it or not, you harbor preconceived notions about risk. The precepts that I will put forth in the ensuing chapters will run counter to at least some of your current understandings. So, a bit of "priming the pump" is required with regard to hopefully convincing you that the methods and definitions that I will pose are logical and universally applicable.

Some Acronyms

I don't know about you, but I find it exceedingly frustrating when reading an article or book to have an acronym used in the text far from where it was defined. You will note, if you haven't already, that I will use certain acronyms quite regularly. I also find it quite vexing, as the author, to repeatedly define an acronym. So, regarding sparsely-used and less-well-known acronyms, if in the upcoming chapters you find yourself wondering just what an acronym means, you can come here to alleviate your perplexity.

CAC = command and control
COF = chance of failure
COS = chance of success
IRR = internal rate of return
MO = matrix organization
NPV = net present value
PM = project manager
RIW = risk-identification workshop
RPP = risk-process proponent
TCOS = total chance of success

A Focus on Success and Value—*Not* on Reducing Uncertainty

If you have had the opportunity to peruse one of my previous books, you know I write these books in such a way that the layman might read them and that the experienced person or expert might benefit. Based on four previous books, my writing style has been described as "folksy," and this book will adhere to that format.

It is not my nature to begin a conversation with a complaint. However, in this case, I just can't help myself. In my risk-related positions, I have run into hordes of folk each of whom would proclaim herself or himself to be an expert in risk. It is not uncommon to discover that their "expertise" in "risk" stems from their experience and prowess in running spreadsheet add-on software that allows them to capture uncertainty by putting ranges around otherwise deterministic values in spreadsheet cells. The view is that "risking" parameters constitutes the practice of putting ranges around deterministic values so that Monte Carlo analysis can be employed. This, to the uninitiated, is the risk process. I am here to tell you that it is not.

The primary goal of most risk analyses should not be concerned with capturing, expressing, or utilizing—as in Monte Carlo analysis—uncertainty. The spreadsheet add-on software packages—and you know which ones I mean—have, in my opinion, done a great disservice to the risk world because they have promoted the idea that risk analysis is mainly about capturing and employing uncertainty. In my classes, I typically state that if I were in a "risk contest" with another individual and I had to give up one aspect of risk analysis in the competition, I would give up the capture and use of uncertainty—and I would still beat the pants off the other contestant. Addressing and reducing uncertainty is nice if you can get it, but it usually should not be the primary focus of most risk analyses.

"Well," you say, "so if uncertainty is not the primary focus of a risk analysis, just what should be the focus?" Funny you should ask because I was just going to address that very issue.

Most credible and comprehensive project-related risk analyses should focus primarily on two things: (1) getting an unbiased view of the COS of the project and/or (2) determining the value of the project. As threat-mitigation and/or opportunity-capture actions are part of any comprehensive risk assessment, a bent toward increasing the COS and toward increasing the value of the project is typical.

Allow me to make a case for the above statements with an example. Let's suppose that you are the person in charge of construction of a new building. The materials-transport and crane configuration that has been foisted upon you by the engineers is one that has been used around the globe for more than 40 years in the construction of, literally, hundreds of buildings like the one for which you are responsible. Decades of statistics clearly show that over a 1-year period—using the "tried and true" construction methods suggested—there is a 35–40% chance (5% uncertainty range) that something of sufficient mass will be dropped from a height that could cause injury to workers or significant damage to infrastructure or other equipment.

Because of your personal concern for workers' safety and due to the company's increasing focus on safety and cost, you are not at all happy with the fact that in 1 year there is a 35–40% probability that something will fall on something or someone with serious consequence. So, you set about redesigning the material-transport equipment and process, the methods used to erect and utilize the crane, and the processes employed by workers to interact with the entire job site.

Your engineers indicate that employing the new techniques and processes will add somewhat to the cost of the operation. However, they estimate that if the new methods are faithfully employed, the probability of serious damage or injury from dropped objects will be reduced to 5–20%. Success in this project is, in part, defined by not causing damage to equipment or harm to workers. The value of the project will be greatly increased if no equipment is damaged and/or if no workers are injured—in spite of the fact that the costs to employ the new methods will be somewhat higher.

So, employing the new methods and processes will increase your chance of success and of the project value. However, your uncertainty has gone up considerably. Forget about the percent probability numbers quoted above. Consider that if you employ the "old" methods and processes, there exist 40 years and hundreds of buildings of evidence that indicate what is the chance of "failure," that is, of a materials drop resulting in serious consequences. You are going to employ new techniques and processes that, if ardently applied, will reduce the probability of a significant drop. However, this being the first time such techniques and processes are being employed, you can't possibly be more sure of the magnitude of the new probability of a serious drop than 40 years of evidence would suggest. Your uncertainty is much higher (you really don't know what the probability of a serious drop will be), but your chance of success (no serious drop) and the value of the project, hopefully, will increase.

Uncertainty reduction is great if you can get it. I am *not* making the case that reduction of uncertainty is a bad thing—quite the opposite. If you can achieve a reduction in uncertainty while simultaneously increasing the chance of success and/or the value of the project, then it is good for you! However, uncertainty reduction should almost always be seen as your "degree of freedom" in any risk analysis. We should typically (but not always) focus any risk analysis on getting the best estimate of success and value as is practically possible. Then, threat-mitigation and opportunity-capture steps should aim to increase the project's chance of success and/or value. If uncertainty has to increase to do this—as in our construction example above—then so be it. If you can achieve uncertainty reduction *and* have an increase in success and/or value simultaneously, then, hey—bonus!

To begin to address the query that is the title of this chapter—"What's the Point?"—I have to focus on success and value. This is a departure from the tenets of most risk-based books in which the main point of attention typically is an entity that is discipline specific such as sales, costs, production, hedging, and the like. The bent of this book will be decidedly different in that it will not focus on any single enterprise type or discipline. Rather, the fundamental precepts put forward here will be as follows:

1. Applicable across a broad spectrum of commercial, academic, governmental, and other endeavors

2. Represent a holistic view of risk, considering threats and opportunities emanating from areas such as law, the environment, commerce, technology, engineering, H&S, finance, logistics, security, and other disciplines

3. Backed up with examples of the translation of a holistic set of risks into impact on the perceived value of projects

Many Contributing Disciplines

Again, addressing the "What's the Point?" topic, it is not unusual for any project in any business or governmental agency to be impacted by threats and opportunities that emanate from a broad spectrum of sources. Figure 1.1 is a generic illustration of this situation. One of the prime lessons of this book will be how to translate the threats and opportunities originating in each of the risk areas, and more, shown in Figure 1.1 into impact on the perceived value of the endeavor. The examples in Figure 1.1 are not meant to be exhaustive—just representative.

Each discipline shown in Figure 1.1—and many more—will have a colloquial view of risk. That is, each area of expertise will utilize tailored verbiage, plots, metrics, and so on when talking about, measuring, managing, and expressing risk.

To begin with, every discipline views risk in a unique manner. I always revert to using the example of how folks in the finance department of a company will view risk very differently than the personnel in the H&S department of the same company. Those in the finance department are charged with getting the best return possible or practical for any investment of company capital. Finance personnel know that low risk leads to low return and that higher risk typically generates greater financial rewards. They, of course, don't want to take on more risk than they can effectively handle (hedge, etc.), but there is no doubt that the prevailing philosophy in the

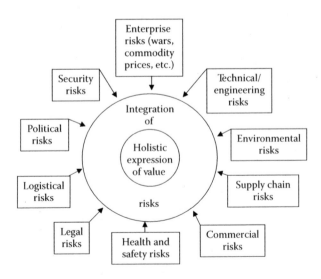

FIGURE 1.1
Subset of discipline-specific risks that have to be integrated to get a holistic view of perceived value for a project.

finance department is that risk is an opportunity. Finance personnel seek risk—to a point—to maximize return on investment.

Contrast this with the personnel responsible for H&S issues in a company. Typically, H&S workers view risks as threats. All risks are to be identified and eradicated, if possible, or at least mitigated to an acceptable level.

Many Views of Risk

Oh, life in the risk world would be so much simpler if it were true that people in different disciplines viewed risk in different ways, but they used the same metric to measure the risk. That is, in a perfect world, the members of the finance department mentioned above and those personnel in the H&S department might see risks as opportunities and threats, respectively, but would employ the same metric to measure risk—such as dollars. Now, I know that expressing H&S risks in dollars is not a socially acceptable practice (so many dollars for loss of a limb, so many dollars for loss of a life, etc.), but in the insurance business and other endeavors, H&S risks typically are monetized. This is how, for example, insurance-premium rates are set.

As I said, it would be great if members of disparate disciplines all utilized the same metric to measure risk, but they don't. While it is true that most risks, ultimately, become translated into some measure of value (money), along the risk-assessment or risk-management journey, various metrics are utilized. Numerous expressions of time, costs, incomes, simple counts, and other metrics are used. But it's even worse than that. Not only do different disciplines view risks in unique ways (like the finance and H&S folks) and utilize different metrics to measure risk, but they employ a wide spectrum of plots, charts, and other methods to express risk. Figure 1.2 shows a few of the very many methods typically employed to express risk.

Project managers, for example, are likely to express risk by utilizing a risk register. This is common because costs, time, mitigation actions, responsible parties, due dates, and so on are important elements of the reward system for the project manager. That is, he or she gets rewarded for getting the project done on time, on budget, and with the allocated personnel. A lawyer, however, is likely to express risk as text. Attorneys find prose to be a natural medium through which they can relate complex issues and concepts. Engineers and modelers are more likely to gravitate toward plots such as cumulative frequency curves. Plots such as the infamous "Boston Square" are favored by project planners. "Traffic light" plots (red, yellow, and green dots) are favored when "risk dashboards" (more on dashboards later) are the aim. And so it goes.

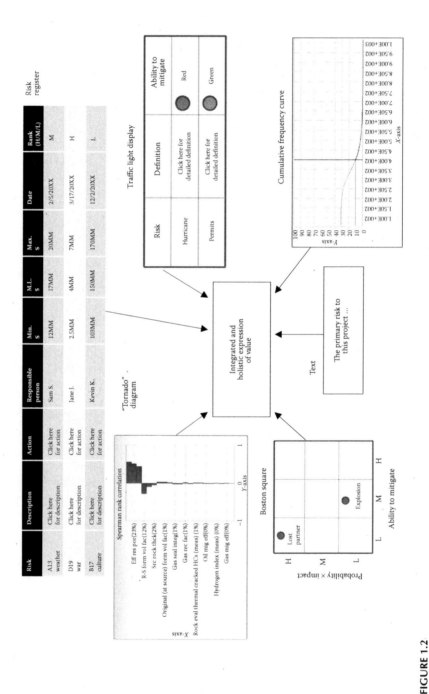

FIGURE 1.2
Example of the expressions of risk that must be integrated to get a holistic view of perceived value for a project.

The trick, then, is to employ a risk proponent—a person—who has the following capabilities:

- Able to speak intelligently with each—or at least, most—of the personnel in each discipline; that is, be cognizant of the colloquial verbiage
- Able to understand how personnel in any given discipline view threats and opportunities
- Able to understand the metrics employed by any discipline
- Able to understand the expression of the metric employed
- Able to translate the utilized expression into a common metric that can be used to impact the perceived value of the project

This, of course, is a tall order. It is folly to believe that any "risk expert" in any given discipline will serve this role. For example, it is not reasonable to expect that the commercial-analyst risk expert will run around the company holding meaningful conversations with staff in the legal, environmental, logistics, political, engineering, financial, security, and other departments. A well-rounded and experienced risk proponent is critical for this task. More about this will be discussed later in this chapter and in this book.

An Aim to Integrate Risks

In the end, it is really all about making decisions. When decision time is upon us, we hope that we can base that decision upon the best information available and upon a metric or set of metrics that are most pertinent to the dilemma. As is my wont, I will relate here a personal example.

Earlier, I served as the head of a corporate risk group. As such, I had responsibility for the usual personnel issues. However, I also had "hands-on" accountability for attempting to ensure that, when corporate decision makers were considering whether or not to fund or sanction a major project, those decision makers were basing their judgment on a pertinent set of metrics. Those metrics had to reflect the impact of the holistic set of threats and opportunities that pertained.

One example is a multibillion dollar construction project in which I was involved. This monumental effort demanded consideration of risks associated with the laws of multiple countries, technology and engineering challenges, commercial hurdles, financial and supply-chain considerations, H&S, as well as security issues, cultural challenges, and many other sources of risk. Paramount among these risks, however, were those stemming from politics. Although a deal with the main political entity had been reached with regard to royalties, taxes, and the like, election time was nigh. A new elected official could change the entire deal. This risk was identified early in the project life, and, you guessed it, as a result of the election a newly elected official changed the terms of the deal, which killed the project that had been in the works for years.

So, although the aforementioned effort was primarily a world-class construction project, it was a very real and potent political issue that caused the project's demise. Why am I making a point of this? I am not a fan of spreadsheets, but the spreadsheet that was used to calculate the value of the project—represented as net present value (NPV)—contained line items typical of any construction-related ledger. Component costs, potential product throughput, yearly operating costs, royalties, taxes, and the like were considered. Missing, however, and not surprisingly, was a line item for politics.

Early in the project's life cycle we had held a risk-identification workshop. How to conduct such a workshop will be discussed later in this book. At that early stage, we identified political threats as some of the paramount risks. However, it was not clear to those who had created the value-calculating spreadsheet why one might wish to incorporate in the spreadsheet analysis the probability and impact of, what seemed to them, an unconventional line item such as politics.

I was not in control of the value-calculating spreadsheet. In spite of my best efforts to persuade the keepers of the sheet to include a specific "soft risk" item representing a political threat (and many other soft risks), I was unsuccessful in being granted such a modification. This caused me to attempt to incorporate the probability and impact of political risk indirectly in other more conventional items in the spreadsheet, such as project timing, costs, revenues, and the like. So, although the consequences of political risk could be reflected in the final perceived value of the project, this indirect accounting for any soft risk is highly unsatisfactory.

Dissatisfaction with this indirect approach stems primarily from the fact that the impact of politics was not specifically addressed; therefore, the consequences of political problems were not obvious by inspection of the spreadsheet input parameters and coefficients. Without direct and explicit representation in the input list, conversations about political threats and about attendant mitigation efforts were less likely to be had. In addition, the probability and impact of a political risk was masked because that likelihood and consequence was expressed only as a modification of an already existing more conventional line item such as a cost or timing.

Again, political problems were identified early as being paramount and, in fact, it was a political threat that killed the project. One might think, therefore, that specific mention and consideration of such a risk might be logical steps in the risk-assessment and risk-management processes. If one thought that, one would be wrong.

It turned out that the traditional processes for calculating value—the format and parameter-input sets—were so ingrained and sacrosanct that modification to include "nontraditional" items such as politics, legal issues, cultural concerns, organizational impediments, and the like was not practically achieved. The folly of this exclusionary practice and the methods for directly accounting for the value impact of a holistic set of risks will be the focus of this book.

Surprise! Integrated Impact Is Almost Always Alarming

You know, a lot of what I am going to relate in this book might seem like common sense. Much of the information conveyed here seems like common sense to me, and, hopefully to you. However, it never ceases to astound me as to how uncommonly these seemingly common-sense practices and concepts are employed.

Sometimes when an author is writing a book, the author falls victim to "writer's block." This is the syndrome of being unable to come up with any meaningful text. With regard to the topic of this section, I have exactly the opposite of writer's block. I have been in the risk business for so long and have seen so many things that I am stifled by the embarrassment of riches about which I could write. Trying to cull the myriad events to select a few that might be related here has been as much a hurdle as writer's block. Thinking back on the very many risk projects with which I have been associated, man, I have seen a lot of silly things!

OK, all seriousness aside, for this section of the chapter to not be a book itself, I had to settle on just a few examples that would represent how integration of the probability and impact of multiple risks can lead to unexpected consequences for project value. Although I could recall a veritable cornucopia of complex examples, such convolute illustrations typically make poor examples for books such as this because the point(s) to be made often are lost in the intricacies of the scenarios. So, I have selected just a couple of representations that are simple and direct to the point.

Dashboards and Traffic Lights

When serving as the primary ramrod for a project risk-assessment or risk-management (RA/M) process, just one of the attendant responsibilities is to interview key representatives from each of the attendant disciplines. It is typical to have to facilitate separate conversations with personnel who represent the legal, engineering, security, commercial, financial, H&S, logistical, and other arenas. In such facilitated interfaces, part of your job is to glean from the other party, among other things, a range of probability and a range of impact associated with each of the threats or opportunities identified in an earlier-held risk-identification workshop.

For the sake of simplicity, let us here assume that each critical discipline identified just one major risk (this really is unrealistic because most disciplines will usually populate a discipline-specific risk register with many threats and opportunities—but please allow me). I related above that I was not a fan of spreadsheets, but my distaste for spreadsheets pales in comparison to my aversion to "risk dashboards."

For those of you who are unfamiliar with such things, risk dashboards are used primarily as a communication tool. Dashboards were deemed necessary because it was considered that decision makers should be shown a very high-level representation of each of the project-critical risk areas. It is not uncommon, therefore, that the integrated impact of risks from an entire discipline (such as commercial, financial, legal, and security) is "boiled down" to a single value or graphic.

Commonly employed on risk dashboards are "traffic light" representations of risk. That is, a red light might represent a "high-risk" discipline, a yellow light an area that contains significant but not project-killing threats, and a green light an arena that represents minimal threat to the project.

Well, the first thing I don't like about such representations is that they focus almost exclusively on threats. Discipline-specific opportunities for the project typically are not considered, or, if they are, such opportunities are not specifically highlighted. It is not so detrimental that any specific discipline represents its overall project-threat level as a red, yellow, or green light, but it is the integration of such representations that is most revealing.

For example, it might be decided that if the threats associated with a given department have a 10% or less chance of causing the project to fail, then that discipline can represent its overall risk to the project with a green light. The folks in the commercial area have decided that there is only a 10% chance that threats emanating from their work could kill the project and, therefore, decide to represent their area's overall risk with a green light on the risk dashboard. Likewise, the people representing the legal, financial, logistical, political, H&S, engineering, security, and other areas have all decided that their areas also represent a 10% or less chance of project failure. Each of these disciplines, therefore, posts a green light on the risk dashboard. A dashboard of such lights is shown in Figure 1.3.

Well, imagine the delight of management when presented with a risk dashboard of nothing but green lights! How could they not imagine that this project will be successful? However, it ought to be obvious to even the most remedial of readers that if each green light stands for a 10% chance of project failure and there are 30 green lights on the risk dashboard, the project is doomed to be a failure. Statistically speaking, at least one of those risks (and in this admittedly concocted case, three of them) will cause failure. When the project fails, nobody can understand why a project with a dashboard of glaringly green lights would have met with an untimely demise.

As I said, a lot of what I will relate in this book would seem to be common sense. It should be common sense that if the project has 30 things that each

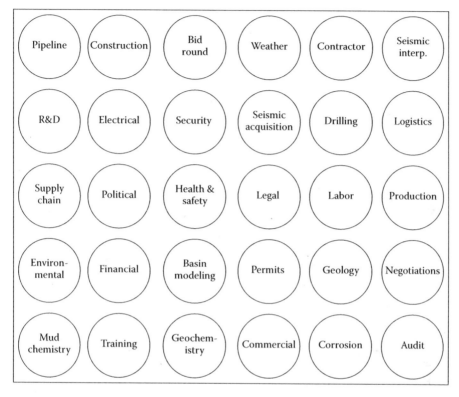

FIGURE 1.3
Example of sources of failure for an energy-company project. Each source is represented by a "traffic light," the color of which is green—indicating in this example a 10% or less chance of failure for that source.

have a 10% chance of killing the project, there is little or no hope of success. It is not so detrimental that any one discipline represents its project-threat level as green. However, it is the integration of all the green lights that ought to boil down to one glaring red light.

Now, you might say that it is unrealistic that each probability of failure is 10% and that a green light can represent 10% or less—and you'd be right. However, it is not uncommon to see dashboards with a couple of yellow lights on them and, maybe even one red light, and to still have the project sanctioned. You might also argue that any dashboard represents a snapshot in time and that disciplines that, say, are represented by yellow lights are working on mitigation actions that should reduce their discipline-specific threat level from yellow to green. That's all very nice, but it has been my experience that follow-through on such mitigation actions is spotty, the effectiveness of such mitigations is overrated, and that reducing the threat level to green still leaves you with the glaringly green dashboard that, if the individual threat levels were integrated, might deserve a red, or at least yellow, color.

I expect that it is still unclear to some of you as to why a dashboard of green lights should lead to failure. Much more about this will be illustrated in Chapter 3.

It's about Time

A second and final example of the impact of risk integration comes from a real and specific project experience in my past. Like the situation described in the previous section, this scenario was relatively simple in nature, involved integration of risks from multiple disciplines and revealed an integrated result that was met with disbelief. Unlike the previous example, this one is about time.

Once again, I was serving as the risk expert on a large, complex, and expensive project. I came to the project after initial risks had been identified and recorded in a risk register (a spreadsheet). Perusal of the risk register revealed that personnel from eight major disciplines had been interviewed and that they had recorded their risks—threats and opportunities—in the register.

Like most risk registers, the areas of impact were broken out primarily as follows:

- Revenue
- Capital costs
- Operating costs
- Schedule

That is, each threat or opportunity listed was assigned to one of these impact areas—the area in which the risk would most likely and most significantly have consequence. It was obvious from even a quick look at the register that more than half of the risks were deemed to have a major impact on schedule—especially start-up time of the project-resultant facility.

Before I joined the project, schedule-impact metrics were collected by someone or by some cadre of people through separate interviews with representatives from each of the eight major disciplines. For reasons that I will dwell upon later, most of the schedule impacts were threats—things that would delay the start-up of the facility. Very few schedule opportunities—things that would result in a sooner-than-expected start-up—were listed.

Few of the schedule threats represented time delays of more than 6 months, and associated probabilities (likelihoods that the schedule threat would actually materialize) were not shockingly high. However, in spite of the many relatively small risks to schedule, the stated start-up date for the project was unchanged from its pre-risk-register date of $N + 2$ years (N being the current year). As stated previously, I have been around the risk world a long time and was immediately suspicious of the expected start-up date.

I took it upon myself to visit again the folks in each discipline who had been interviewed before—to create the risk register. I did not attempt to capture again information that had already been recorded. Rather, I simply asked a few questions regarding what resources would be drawn upon to enact the mitigation plan(s) associated with each schedule threat or the resources necessary to capture each schedule opportunity.

Having collected this information, I then knew which threat-mitigation or opportunity-capture actions could be implemented in parallel (i.e., at the same time). For example, if a commercial threat required that 100% of the legal team's time be devoted to mitigating the problem, then another threat or opportunity that also required the attention of the legal team could not simultaneously be addressed.

Utilizing this information and that already recorded in the risk register, I built a Monte Carlo model (I will specifically address Monte Carlo analysis in a later chapter). The model probabilistically calculated the facility start-up time—accounting for those threat-mitigation and opportunity-capture actions that could happen in parallel and those that would have to happen in series (one after the other). Remember, the original facility start-up year was $N + 2$, where N was the current year. The resulting histogram of start-up years (see Figure 1.4) did not even display year $N + 2$. The first year of possible start-up—and not a very good chance of starting in that year—was $N + 5$. Year $N + 7$ represented the tallest bar on the graph.

When I showed the resulting start-up histogram (Figure 1.4) to project managers, they were sure that I was in error and accused me of being so. In an attempt to convince them of the accuracy of the prediction, I created a relatively easy-to-understand explanation of the logic of the Monte Carlo model. In a meeting with all of the decision makers, I presented the raw data from the risk register and the Monte Carlo logic for combining the raw data.

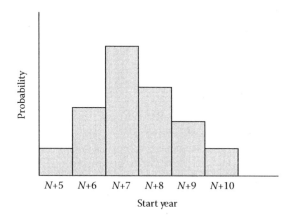

FIGURE 1.4
Bar chart of projected project-start years resulting from the integration of risks and risk-mitigation or capture actions. "N" is the current year.

In the end, the decision makers accepted the logic of the model, but they still did not accept the projected start-up date—now preferring to suspect the credibility of the raw data (probabilities and schedule impacts associated with individual threats and opportunities). A subsequent review of the raw data did not significantly alter the original metrics.

OK, Glenn, so what's your point? My point is that although all of the schedule-risk probability and impact data had been collected and recorded and that each individual schedule threat did not probabilistically amount to much, the integrated impact of the seemingly small schedule threats was exceedingly significant. It was unlikely that perusal of the risk register by the untrained eye would lead to suspicion that the original start-up date estimate was woefully optimistic. In the end, project management could not avoid the realization that unless something changed with regard to threat mitigation and opportunity capture, there was no real chance that the facility would start up when required for the project NPV to be viable.

Now, I don't actually believe that the two relatively simple examples described above would convince even a mildly skeptical reader that the integrated impact of multiple risks can be perceived very differently from the presentation of individual stand-alone risk probabilities and consequences. However, it is nonetheless true that because the integrated impact of risks can be exceedingly divergent from the perceived impact of individual and nonintegrated risks, it is a very good practice to always attempt to convert all major threats and opportunities to impact on value and to integrate those impacts.

For Absolute and Relative Purposes

As I stated previously, some of the ideas, concepts, and processes about which I will write should seem to be common sense. This issue, it seems to me, should fall into that category. However, I am often amazed at the failure, on the part of a project team, to recognize the difference between estimating the value of a project for the purpose of project ranking versus calculating the absolute value of the undertaking. I will relate much more on this issue in ensuing sections of this chapter in which I explain in detail the expected value of success (EVS) and the expected value for the portfolio (EVP), but this subject warrants a preview here.

In my previous work (see *Risk Assessment and Decision Making in Business and Industry—A Practical Guide,* 2nd Edition by Koller, 2005), I have written extensively about the importance of establishing "What is the question?" for any given effort. That is, before any effort is expended on a risk-assessment or risk-management process, individuals critical to the success of the risk analysis need to come to consensus regarding just what problem they are attempting to address.

For example, it is not uncommon in the very early stages of any project life cycle to hold a risk-identification and risk-ranking workshop. Much more on the method and format of this exercise will be related in subsequent chapters of this book. At such a meeting, representatives of the critical disciplines convene to brainstorm the major risks to the project.

Before this can effectively happen, however, it must be established just what problem the risk analysis is aimed at solving. If such a workshop were convened to address, for example, the threats and opportunities associated with the construction of a new factory, it would be typical to have in attendance representatives from the commercial, financial, construction, supply chain, legal, management, engineering, and other areas. Even though each of these discipline experts would collectively swear that they were all there to address the same problem—the construction of the new factory—in reality each individual has a very different view of the critical issue to be resolved.

At such meetings, I often employ the tactic of asking each person in attendance to take out a piece of paper and a pencil. I then ask them to imagine that the risk analysis is complete and that they are standing by the printer that is now producing a piece of paper on which is their perfect answer to the problem we are assembled to address. I ask each of them to then draw on his or her piece of paper just what is issuing forth from the printer. Put differently, I ask each of them to draw on the piece of paper just what is the concept of the perfect output from the risk analysis.

Usually, there was much consternation and scratching of heads, but in the end, each participant hangs his or her rendition of the solution on the wall. It never ceases to amaze me how different are the imagined solutions.

Engineers typically will have scribed some sort of plot that is technical and quantitative in nature. A commercial analyst might also draw a plot that, not surprisingly, will have at least one axis that relates to money. A project manager will ordinarily generate some sort of Gant chart that relates his or her angst about budget and timing. And so it goes. If everyone is not to be disappointed by the actual outcome and output from the risk analysis, it is critical at the outset to bring the group to consensus regarding just what we expect the risk effort to address and what output it will produce.

You might be justified in now asking, "OK, Glenn, what does this have to do with the subject of this chapter?" I would respond to that question by harkening back to a statement in a previous section: in the end, it is really all about making decisions. Given that a primary focus of this book is value, I postulate here that it is essential—early in the risk analysis—to determine and agree upon just what value metric is desired.

Although a more detailed delineation of this issue will be related later in this book, it is necessary that I here address the issue of "relative value" versus "absolute value." To accomplish this, I will make use of a very simple (overly simple, you might argue) example.

Let us suppose that you are the manager of a project to build a new factory. The factory will produce widgets. The technology for widget production to be utilized in the new widget plant is revolutionary and untried. Hence, there exists, let's say for the sake of this simple example, a 50% chance that the new technology will fail and that no revenue will be realized from the new factory.

If the factory is successfully constructed and operated, it is projected to produce 500 widgets per year. The combined output from the other five widget plants that all employ the old and time-tested technology is also 500 widgets.

Imagine that you are approached by the person from corporate headquarters who is charged with making a forecast for corporate widget production for next year. This person asks you to predict your contribution to next year's production. You cannot assume success of the project because it is well documented that there is a 50% chance that you will produce nothing. Thus, you would deliver to the corporate forecaster a potential plant output of 250 widgets (500 damped by the 50% chance of failure). If each person queried by the forecaster delivers to him a probability-weighted production value, then the sum of those probability-impacted metrics should be a good approximation of next year's actual performance—allowing for the fact that the "rolled up" value represents a weighted mean.

This probability-weighted value is also the metric that would and should be used by corporate portfolio managers to compare your plant project with other investment opportunities. Therefore, the 250 widgets and the resulting revenues would represent the relative value of your project in the corporate portfolio.

Now, suppose that the manager in charge of logistics visits you and says, "Hey, I need to plan for the number of train cars I will need next year to transport your widgets to market. How many train cars do you recommend I lease for this project?" In this case, you have to plan for success. If the venture fails, then losses will be taken across the board—including logistics. The possibility and magnitude of such losses have to be quantitatively and qualitatively considered before giving the "OK" to go ahead with the new factory. If the specter of such losses is catastrophic, then the project likely should not be sanctioned.

In the case of the forecast for train cars, you have to plan for success. If you considered the 50% chance of failure in your train car forecast and told the logistics folks to plan for half the number of cars you would need if successful and the plant does succeed, there would be insufficient train-car capacity to transport the widgets to market, and the economics of an otherwise successful venture likely would be ruined. So, if it will take N cars to transport the 500 widgets to market, you tell the logistics representative that you will need N cars. The cost of N represents an absolute value—not impacted by probability—to be utilized in planning.

So, with this grossly oversimplified example, I have described here the need for and use of both absolute and relative measures of value. When undertaking any risk-assessment or risk-management project, it is essential that very early in the process the project team comes to consensus regarding the type or types of value measures that will be required. Collection of data, the relevant disciplines, the construction of the risk model, and many other factors will be critically impacted by the ultimate use of the value metric(s). Plan ahead!

Where We Would Like to Be

Given that the title of this chapter is "What's the Point," I will postulate that that part of "the point" is to describe where we would like to end up with regard to risk monetization effort that successfully estimates value. Such an achievement would be composed of the following characteristics.

Holistic Set of Integrated Risks

A credible job of projecting value or probability of success can only be accomplished if all pertinent threats and opportunities are considered. In a previous section of this chapter, I related my experience as the main risk advisor on a multibillion-dollar construction project. The spreadsheet used (and, again, I am not a fan of spreadsheets) to calculate the NPV of the project contained all of the typical parameters such as commodity prices, production capacity, revenues, fixed costs, operating costs, taxes, royalties, and the like. Nowhere was there a line item for politics, cultural issues, organizational hurdles, and other "soft" risks. Although I was successful in capturing soft risks in the project risk register, I was unsuccessful in convincing the financial-types who controlled the value-calculating spreadsheet to include specific line items for these critical considerations.

Consequences of these soft-risk parameters had to be incorporated—and essentially hidden—in the values entered for items that were represented as line items. This is an exceedingly poor way to approach an assessment. This is especially true given the fact that it was, in the end, a foreseen and documented political threat that killed the project.

So, as illustrated in Figure 1.1, it is not only essential to consider threats and opportunities from all possible sources—a holistic set of risks—but it is critical that such risks be overtly represented in any value-forecasting process. It is not nearly sufficient to consider in the risk register threats and opportunities emanating from legal, commercial, logistical, political, financial, security, H&S, engineering, technical, and other sources. It is best practice to explicitly include these risks in the value calculation and to not hide them in the impact of traditionally considered parameters. More on how these nontraditional risks are explicitly incorporated later in this book. I have in a previous section of this chapter already beat you about the face and head regarding why it is essential to consider the impact of an integrated holistic set of risks.

Threats and Opportunities Translated into Impact

Thus far, I have harped on the fact that nontraditional risks need to be considered and their probabilistic impact reflected in the estimate of value. This is equivalent to saying, "Once you have gathered all the risks, a miracle happens and the impacts are realized." Performing the miracle can be tricky.

Over many years, I have had the privilege to have close interactions with several universities and have taught courses at those institutions. With respect to risk assessment and risk management, it is my considered opinion that our educational establishments do a poor job of preparing students for the real world in which risks have to be considered.

Figure 1.2 only begins to illustrate this point. It is folly to imagine that, just for example, a lawyer will view or express risk in the same manner as will an engineer. Representatives of the legal profession will tend to express their view of risk as some form of text. In Figure 1.2, this is represented by the verbiage in the lower left portion of the figure. Engineers, in stark contrast, tend toward quantitative expressions of risk as exemplified by the cumulative frequency plot in Figure 1.2.

If expression of risk was the only complicating factor, then finding a way to translate each of these different expressions of risk to a common format or metric might be relatively simple. However, the challenge is considerably more diabolical.

It is NOT that representatives of different disciplines are viewing or experiencing threats and opportunities in the same way and simply expressing those views and opportunities in an unharmonious manner—oh no, they are viewing and experiencing very different things that they then express in a colloquial and unique way.

An attorney might see a threat from a competing company that could materialize as a patent on technology—by the competitor—that would preclude the attorney's company from future utilization of already-imbedded technology that is crucial to the company's profitability. Advice from the lawyer might come as a letter or memorandum that recommends the immediate filing of a patent on the technology that would allow the attorney's corporation unfettered use of the technology in the future.

An engineer at an energy company might see an opportunity to increase production in an oil field by introducing secondary-recovery technologies. While such methodologies exist, their effectiveness can be spotty—that is, implementation of the technologies can result in huge investments that result in revenues that don't come close to offsetting the costs. Therefore, this opportunity is a risk—an event that has an uncertain effectiveness. The engineer is likely to express the possible outcomes of the endeavor by producing cumulative frequency plots—like the one shown in Figure 1.2—which relate possible outcomes to likelihood.

As illustrated in Figure 1.1, if a most clear estimate of perceived project value is to be attained, then the threats and opportunities from all pertinent disciplines have to be considered. Figure 1.2 illuminates the fact that each of those disciplines can view and express risk in very different ways. Integration of all of the views and expressions is the trick—and is the main subject of this book. So, what does it take to be able to do this?

Being Able to Talk to People of Various Disciplines

Often, it is a single person who spearheads the risk effort in a project or in a corporation. Sometimes, the task falls to more than one person, but that situation is relatively rare—mainly due to lack of expertise and resource constraints.

It is not likely that a junior member of the staff will be able to fill the role of risk proponent—that is, the role of the person who has to "pull it all together" as illustrated in Figure 1.2. The person responsible for this task has to have at least a rudimentary understanding of each pertinent discipline and be able to carry on a meaningful and effective conversation with representatives from each discipline. For example, it is unlikely that in a conversation with a very busy engineer that the engineer will "couch his or her terms" when expressing their view of the threats and opportunities that pertain. The recipient (risk proponent) of the conveyed information has to have sufficient background in the field to at least be able to comprehend the engineer's expression of risk—which would undoubtedly include much engineering-specific jargon. It is only through experience that the risk proponent, who might not be an engineer, can gain such proficiency. And so it is with other disciplines. Being able to intelligently converse, or at least listen to, the expressions of experts in a wide range of disciplines is a rare talent.

Being Able to Translate That Expression of Risk into Something That Can Impact Value

Listening to and comprehending the expression of risk by a discipline expert is just the first step in the risk-monetization process. A next crucial action is to be able to translate that articulation of risk into a metric that can be used to impact value.

Consider Figure 1.5 in which an environmental engineer has used a traffic light method (red light = high threat, yellow light = moderate threat, and green light = low threat) to express the likelihood and impact of a environmental risk. Using the process outlined in Figure 1.5—which is admittedly a grossly oversimplified representation of real life—it can be seen that we can translate a red dot into a financial metric that then might be utilized in a value-calculating set of equations. Such translation of each of the risk expressions in Figure 1.2 (and many more) is a skill that is required for successful risk monetization. Translation is again addressed in the "Translated Impact—Common Unit" section of Chapter 2, in the "Translation of Risks" section of Chapter 5, and in the "Create Translation Table" and "Use Translation Table to Convert to Common Measure" sections of Chapter 6.

Integrated into "The Way We Work"

Success of a risk process mainly is a cultural issue. If such a method is viewed as an "add on" to already-existing responsibilities, then resistance can be

Red traffic light
associated with an
environmental risk

Steps to translate a "traffic light" into a quantitative measure of risk:
  1. Discuss risk with discipline expert.
  2. Collect ranges for probability and two levels of impact.
  3. Create a metric that can be inserted into the project economic model.

Example:
  What is the probability range that we will realize this environmental threat?
    Minimum, peak, and maximum values.
  What is the range of days we will be delayed if we realize this environmental threat?
    Minimum, peak, and maximum values.
  What is the range of dollars/day if the environmental threat materializes?
    Minimum, peak, and maximum values.

Environmental threat cost range = (prob. range) × (delay range) × ($/day range)

FIGURE 1.5
Example of quantifying a red ("high risk") traffic light, so the risk impact and probability can
be used in value-assessing calculations.

staunch. A more successful approach is to truly understand the already-
established workflows of each affected discipline.

For example, if an engineer is told that at the end of his or her normal list
of tasks, he or she now has to add on the task of a risk assessment, unless
there is specific reward to doing so, such a risk analysis is likely to follow
the letter of the law ("check-the-box" mentality) than the spirit. A more
fruitful approach is to gain an in-depth understanding of the already-
established workflow of the engineer and to incorporate elements of the
risk process into that stream. In that way, when the engineer (or representative of any discipline) comes to the end of the assigned and accepted set
of tasks, the elements necessary for contribution to the risk-assessment or
risk-management processes will be at hand.

We'd Like to Keep Two Sets of Books—The EVS and EVP

Again, this is a cultural issue. It should be accepted practice that each risk-
impacted endeavor be represented by two measures of value—the EVS and
the EVP.

In the "For Absolute and Relative Purposes" section, I employ an exceedingly simple example of a widget-producing company that will introduce
a new and untested-in-full-production-mode widget-producing technology.
This new method is estimated to have a 50% chance of abject failure—that is,
the new plant will fail to produce any widgets at all.

I postulate that when the logistics folks ask the new plant manager how many train cars he or she will need to transport the widgets, that manager has to plan for success. If the new plant is successful, it will produce 500 widgets a year. Therefore, the EVS—the value the project team plans around—is 500. Planning for anything less would significantly damage or ruin the value of a successful operation.

However, when the gent from the corporation shows up and says that he needs an estimate of widget production from the new plant so that he can roll that number into a corporate-wide widget-production forecast, the chance of failure (50%) has to come into play. The EVP has to be the probability-impacted value of 250. The value of 250 (and the associated revenues) would also be the value that corporate planners use to compare this project to other projects in their corporate portfolio of potential investment opportunities.

So, we would like to be in a place where it is accepted practice that each project is represented by two values—the EVS and the EVP. As logical as this might appear to be, it is far from being universally embraced methodology.

Why, you say, is this practice not standard? There are several very practical reasons. First, most companies have their financial and other practices reviewed by external auditing firms. It is just simply true that most of these auditing firms do not appreciate the fact that any project can be represented by two distinct values. This is just not standard practice. And, heaven forbid, the accepted metric for project value be expressed as a probabilistic range! If the project value is not expressed as a deterministic measure (i.e., a single value), then the very fabric of society unravels. So, although such measures as the EVS and EVP are exceedingly useful within a company, governmental agency, or academic entity, the audit world has some catching up to do.

Another practical reason that the EVS or EVP concept is slow to catch on is that it is perceived to require more work. While this is not necessarily so, any attempt to implement new and different methodologies that even hint at a greater burden is likely to be met with real and persistent resistance.

Organizations That See the Benefit of Doing This—The Incentive System

This subject was the primary focus of my entire book *Modern Corporate Risk Management—A Blueprint for Positive Change and Effectiveness* (see Reference section at the end of this chapter), and I will much more expound on this subject later in this book. However, because it is a salient element of any successful risk-assessment or risk-management effort, it warrants at least passing mention here.

For reasons that will be divulged later, it is absolutely essential that those upon whom the risk process is foisted believe that adherence to and implementation of such a method will work in their best interest. Rationales for resisting the enactment of a risk process are many and valid.

A well-thought-through (so that unintended consequences don't torpedo the effort) incentive system that rewards compliance is required. The benefits of alliance with the risk process have to be proclaimed and, more importantly, have to be true.

The incentive (reward) system typically works against risk-process implementation. For example, any long-term project usually starts with a cadre of bodies who deal with the initial concept ("Is this a good idea?"). If the project is deemed credible, then it is passed to those who will outline the scope of the project and will generate design options. After that, the project will pass to those who will select a design and detail the process of how the project will be executed. The construction or implementation crew then inherits the project to actually get it "built." Yet another separate group of employees ultimately inherits the project for operation.

In any credible risk process, a risk-identification workshop is held—at least—at the beginning of the project's life. In the example above, this would be during the tenure of the group charged with deciding whether the project concept is one that might be practical and profitable. In that risk-identification workshop, risks to the project throughout its entire life are brainstormed. Many of the risks defined will be those that will occur "down the road" during, for example, the design, or construction, or operation phase.

Along with each risk, a threat-mitigation or opportunity-capture action is identified. That is, what step (mitigation action) will the company take now to cause the threat to never materialize? Conversely, what step (opportunity capture) can the company take to attempt to assure the capture of an opportunity sometime down the road?

Remember that this risk-identification workshop is being held early in the project lifetime—in this case, during the project-concept-judgment phase. The person in charge of this phase gets rewarded for getting his or her part of the job done on time and on budget. If several million dollars worth of mitigation and capture actions are identified, what incentive does this person have to spend the money and time to execute steps to mitigate threats and capture opportunities that might or might not happen, and, that certainly will not happen "on his or her watch"? It's a pretty tough sell to convince that person or group that they should spend money and time for the sake of a better project (take one for the team, so to speak) when those actions will likely push them over budget, cause them to miss their time benchmark, or both. Their incentive system does not reward such behavior.

The bottom line is that incentive systems in most organizations are not set up to favor the risk-process approach. Unless organizational changes can be made, implementation of the most sophisticated and technologically advanced risk process has little chance of becoming self-sustaining in the organization. So, the first thing to be done if one is attempting to implement a risk process is to survey the organizational structure and the incentive system. If it is deemed that the incentive system runs counter to that which would be required to implement a risk process, then changes in organizational arrangement need to be first addressed. If a risk-process-accepting organization (incentive system) can't be created, then you are, indeed, wasting your time in trying to implement a far-reaching risk process.

Am I Wasting My Time?

So, if you are currently a risk proponent within a company or you are considering such a post, the question, "Am I wasting my time?" might occur to you before you start the journey of risk-process implementation. I know that this might surprise you, but my answer is a qualified "yes."

Far too many risk processes initiated within corporate entities focus too much on the technical aspects of the risk process. While technology is important, it pales in significance to the cultural, political, and organizational aspects of process implementation. If these aspects of foisting a risk process upon any organization are not carefully considered, it usually is true that the risk-process proponent is, in fact, wasting his or her time. I have published an entire book that addresses the challenges of implementing any risk process in modern organizations: *Modern Corporate Risk Management: A Blueprint for Positive Change and Effectiveness.* I will relate here just a few of the hurdles to risk-process implementation.

Imposing a risk methodology on any organization means threat and change. Perceived threats include the perception by some people that the process they already use will be changed or abandoned, that a person's or group's influence will be diminished, that the new process will impose radical and challenging hurdles for advancement, and any number of other perceptions. Change itself is typically reason enough to generate "push back" from the organization. If it is not abundantly and crystal clear why the risk process is absolutely necessary (if it ain't broke, don't fix it), then a change to a risk-based process will not be welcome. Change usually means abandoning a familiar and comfortable process for one that is unknown. This usually is a source of resistance.

Support from the Top—Beware of What You Ask for

Another reason why you might be wasting your time in any effort to bring a risk-based methodology to an organization is lack of support from "the top of the house." If upper management—at the highest levels—does not stand behind the new process, then the probability of implementation success is greatly diminished.

Upper management needs to convey to the organization the following:

- The critical reason why the risk process is necessary.
- Those who embrace the new process will realize benefits.
- It is expected that everyone involved will cooperate.

- Training will be provided.
- Benchmarks will be set and progress will be communicated to the organization.

Even with this, it is not atypical to find that the organization adheres to the letter of the law, but not to the spirit of the law (Yeah, we held the required risk review between 8:00 am and 8:10 am last Thursday .... A check-the-box mentality can quickly result if the affected employees do not truly believe in the benefits of the new process.

It is also necessary to be prepared for the onslaught of cries for training and help that will result from the risk-process implementation. I almost experience this firsthand. Let me relate a true and real-life story.

I worked for a major corporation in the early 1990s. I had been advocating the implementation of a risk process that would impact much of the corporation. I knew and occasionally chatted with the President of the corporation. Over time, he came to see the benefits associated with the risk process and he arranged for me to have two presentations to the CEO.

At the end of the second of those presentations, the CEO said—and I paraphrase: "OK, I'm convinced. Should I put out a statement indicating that the organization should begin to utilize this risk process?" At that very moment, I realized that I had gotten what I had asked for, and I really wasn't prepared to deal with it. Like the dog chasing the car—*OK buddy you caught it, now what?*

I suddenly came to the realization that if the CEO issued any sort of edict, I would immediately be flooded with requests to deliver the technology, train an incalculable number of people all over the world, hold innumerable hands while the people tried to implement the process, and so on. I had neither the financing nor the time nor the staff to back up the granting of my request.

It certainly is a chicken-and-egg situation. I could not "gear up" the staff and other resources before I had gotten the OK from the CEO to go ahead with risk-process implementation. However, if I got the OK and an edict to the organization, I could not ramp up resources in any reasonable time to be effective. So, I had to decline the offer of a CEO-issued edict and try to implement the process on a piecemeal basis.

Loss of credibility is just one result of trying to implement a process that you are ill suited to support. Worse is the resulting attitude of the organization toward risk processes: "Yeah, we tried that silly thing once, and it didn't work." Worse yet is the opening of the corporate doors to all sorts of charlatans who would claim to be risk experts but could not spell risk the day before. When the various parts of the organization realize that they can't get the help they need from you, they will resort to external experts some of whom are inept and some of whom are qualified, but bring to the company a philosophy and methodology that is not in line with the process that you are attempting to foist on the organization. All of these things make it better to have no risk process at all. So, if you don't have in hand the resources to do a quality implementation of your proposed risk method, then you truly are wasting your time.

Communication

As stated previously, in my book *Modern Corporate Risk Management: A Blueprint for Positive Chance and Effectiveness,* I address nearly all of the challenges to risk-process implementation and offer practical advice regarding how to address each issue. Of course, given that the subject of this book is the monetization of risks, this text is not the place to rehash all of the concerns. However, I will mention here yet another major challenge—communication.

At first, communication issues might sound and seem mundane—even trivial. I can tell you from first-hand experience that the hurdles erected due to communication foibles are among the most daunting and rank as some of the most vexing to remedy.

Just like the finance and H&S departments—mentioned in a previous section—which respectively view risks as opportunities and threats, most disciplines within any organization will have developed colloquial risk-related verbiage. It is sufficient enough a challenge to aspire to persuade personnel in different disciplines to utilize the same terms. It is doubly difficult to ensure that, even though denizens of disparate disciplines are using the same terms, they mean the same thing when they use those terms.

Before implementing a broad-based and holistic-risk process in an organization, if you polled the populace regarding the meaning of the term risk, it is likely that you would record as many unique definitions as people polled. Most folks would, in some way, relate risk to the downside—that is, to threats. However, it becomes abundantly clear when we focus on value (i.e., money) that this attitude falls short.

For example, if I were to ask you to make a probabilistic estimate of your personal take home (i.e., net) income for the upcoming fiscal year, you might consider the "downside" impacts such as taxes, contributions to savings plans, automatically deducted payments for loans, payments for healthcare coverage, and the like. These are all "threats" to the net income. However, you would be remiss in your duties if you did not consider the "upside" impacts or opportunities for greater net income. You might get a cost-of-living raise next year (which you did not get this year because of hard economic times), you might realize a bonus in the coming year, you might be promoted which comes with a salary boost, and so on.

So, what's the point? The point is that when considering probabilistic events (risks) that can impact an economic forecast (and there is nothing unique about this economic example—this logic can be applied nearly universally), both threats and opportunities have to be considered if you are going to be able to generate a credible net-income forecast.

Even in the H&S arena where risks are traditionally viewed as threats, when estimating costs related to injuries, it is valid to consider not only the rate at which accidents happen and their financial consequences, but a fore-cast of H&S costs must include potential opportunities such as the adoption of new procedures, the implementation of new and safer equipment, new training programs, and the like.

It can be a real challenge to entice those who have strong traditional views of risk to change their mindset. It is essential that this metamorphosis tran-spire, however, if best estimates of risk-impacted probabilities of success and value are to be had. You would not like to be in the situation of com-paring the values and chances of success of two projects one of which is impacted by only the threats and the other which has been influenced by both threats and opportunities. The independent evaluators of both projects could both claim to have performed a risk analysis of the projects, and from their very different points of view, they would both be truthful. This is not what you want.

Another glaring example of risk-related miscommunication having a sig-nificant impact on the fortunes of an organization relates to the meaning of the term "uncertainty." The definition of this term will be addressed in gory detail in Chapter 3, but I would like to take this opportunity to pique your interest.

As I will harp on multiple times in this book, I believe that some of the spreadsheet add-on software programs have done a disservice to risk assess-ment because they convince users—overtly, or by implication—that risk is all about uncertainty, which it is not. Proponents of those software packages focus on putting ranges around otherwise discrete values ("capturing uncer-tainty") as a preamble for Monte Carlo analysis. Please don't fall for this ploy.

Consideration of uncertainty is a sticky wicket, so to speak. Because defini-tion of terms will be addressed in detail in Chapter 3, I will here only set the stage for why it is essential to the art of communication to be clear about the definition of terms such as "uncertainty."

Suppose we resort to the tried and true example of the coin flip. We know that the probability of getting a "heads" is 50%. There is no range around that probability—it is a deterministic (i.e., single valued) likelihood. Now, I ask you, on the next coin flip, are you certain of getting a "heads"? No, you are uncertain. You are uncertain even though there is no range around 50. Is this the uncertainty we mean when we say "uncertainty"? Before you answer, consider the next section.

Imagine that you just dropped to Earth from Plothar (a technologically advanced, but well-meaning planet). You are not familiar with coin flips but are shown a coin and asked to estimate the probability that a flip will result in a "heads." This flipping thing is considered uncouth on Plothar where the concept of money does not include physical manifestations. Nonetheless, you are asked to estimate the probability of a coin-flip outcome. In an attempt to be accommodating, you offer an estimate of 40–60%—a range. Is this what we mean by "uncertainty?" This is different from the deterministic example in the preceding paragraph. Hmmmm....

If it is your job to collect probabilities and consequences related to threats and opportunities, then you had better be sure of what you are attempting to determine regarding uncertainty associated with those likelihoods and impacts. Believe me, people are thinking unique things when you query them regarding uncertainty. Coming to agreement—early in the risk process—as to the meaning of the term will hopefully avert the situation of coming to the end of your interviews and realizing that the "uncertainties" expressed by individuals can't be consistently utilized in the analysis because those expressions have little common ground.

These are just a few examples of communication issues. It is critical that communication becomes a main focus of any effort to implement a risk process in an organization. I will elaborate on this subject in later chapters.

Can't Afford to Implement a Risk Process

Of course, the gist of this section—given its title—is going to be, "You can't afford *not* to implement a risk process." I suspect you saw that coming.

There are several primary sources of financially based resistance when it comes to applying risk principles to any project. The less insidious is the cost–benefit argument. That is, the costs associated with threat-mitigation and/or opportunity-capture actions relative to the probability-weighted financial gains is either not believed, or, is not overwhelmingly capacious or obvious.

As we shall see in Chapter 3, risks by definition are probabilistic events. That is, they might or might not materialize. Therefore, any cost associated with a threat, for example, is something that is not sure to impact the project. Early in the project life one should hold a risk-identification workshop at which project-life-long threats and opportunities are identified, are succinctly described, are "valued" with regard to their impact on the project if they happen, are assigned a probability of happening, and have described for them the mitigation or capture actions and associated costs.

Mitigation or capture costs are relatively concrete. I return here to the simple example of the "project" of taking a long trip in a car. One of the threats identified at the pretrip risk-identification workshop is that of a wheel coming off the car while driving. The discovery of loose lug nuts has lead us to discovery and definition of this threat. The mitigation action is, before the trip, to take the car to the local garage and, for $20 have the lug nuts tightened on that wheel and on the other three.

It is obvious to even the most remedial of readers that the mitigation cost is $20. Damage estimates associated with wheel loss, however, are less obvious. Depending on the speed and attitude of the car, the number of other vehicles near the car in question, the type of road on which the wheel loss occurs, and other factors have lead to damage estimates of between $300 and $2500. Further, the probability of the wheel coming off the car is not "a sure thing." That is, the probability is not 100%. Estimates of probability of wheel loss range from 0% (the wheel does not come off) to 30%.

I know that most readers are now mentally applying the 30% to the estimated repair costs ("Let's see, 0.3 times $300 …") and that's fine, but it's not
the point. *So, what is the point?* The point is that there does exist the possibility
that this problem might not arise at all. In addition, it is not uncommon for
people in a risk-identification workshop to discuss or argue for significant
lengths of time about probability ranges, but in the end, not really believe the
percent values that result from that interaction.

So, it is the very real $20 that would have to come out of your pocket right
now versus a cost that might not happen, or, a cost that has associated with
is a weighting factor (the probability that it will happen) that, in the mind
of the person who has to pay the bills, lacks credibility mainly because they
want it to lack credibility. Remember, the person in charge of the project in
early days is working under an incentive system that rewards him or her for
getting their part of the project done on time an on budget. If the risk being
considered is "down the road" (no pun intended considering the cart-trip
example), that $20 is really easy to ignore.

A more nasty reason for making the argument that "We can't afford to
implement a risk process …" is that of the organizational cost of implementation relative to the status quo. It is typical to hear arguments that center around the number of man hours that will have to be spent in doing a
credible job of risk assessment and management relative to the current cost
structure. Labor costs, material costs, training costs, procedure-change costs
(i.e., revamping the way they do business) etc. are all valid considerations.
All I will say in this early section of the book is that the proponent of the
risk process needs to be prepared to present credible, believable, and—most
importantly—practical evidence that the financial benefits associated with
risk-process implementation more than compensate for the costs.

Of course, the best way to make that argument is to design the risk process
so that it is not perceived as an "add on" but is cleverly constructed to appear
as an integrated part of their already-existing workflow. This means that the
risk proponent must have an intimate working knowledge of the business
into which he or she wishes to inject a risk-assessment or risk-management
method. Do your homework!

A final point I will mention here, but by no means the last hurdle in the
real world, is the fact that financial benefits to projects from risk processes are
typically not immediately evident. In fact, on a single-project basis, it might,
in fact, be true that a risk process might cost more than the reward. However,
when applied to the entire portfolio of projects, there is no question that the
relative low cost of mitigating threats and capturing opportunities yields
a significantly more profitable portfolio. The lack of a direct line of sight
between process implementation and benefit can be a powerful "argument"
for not implementing a risk process. Again, the risk proponent should have
on hand incontrovertible proof of benefits to projects just like the one he or
she is attempting to impact. That is, showing statistics about how implementation of a risk process has reaped great benefit in the car-building business

when standing in front of a pharmaceutical group is not likely going to be perceived as a compelling argument.

Suggested Readings

In addition to books and articles cited in the preceding text, this list includes other material that the reader might find helpful and relevant to the subject matter discussed in this section of the book.

Argenti, P. A. *Corporate Communication.* 2nd ed. New York: McGraw-Hill/Irwin, 1997.

Bartlett, J., et al. *Project Risk Analysis and Management Guide.* 2nd ed. Buckinghamshire, UK: APM Publishing Limited, 2004.

Koller, G. R. *Risk Assessment and Decision Making in Business and Industry: A Practical Guide.* 2nd ed. Boca Raton, FL: Chapman & Hall/CRC Press, 2005.

Koller, G. R. *Modern Corporate Risk Management: A Blueprint for Positive Change and Effectiveness.* Fort Lauderdale, FL: J. Ross Publishing, 2007.

Lam, J. *Enterprise Risk Management: From Incentives to Controls.* New York: John Wiley & Sons, 2003.

Mulcahy, R. *Risk Management Tricks of the Trade + PMI-RMP Exam Prep Guide.* 2nd ed. Minnetonka, MN: RMC Publications, 2010.

Pritchard, C. L. *Risk Management: Concepts and Guidance.* Arlington, VA: ESI International, 2001.

Schuyler, J. R. *Risk and Decision Analysis in Projects.* Newton Square, PA: Project Management Institute, 2001.

Skinner, D. C. *Introduction to Decision Analysis: Beginning Coursebook.* Gainesville, FL: Probabilistic Press, 1996.

2

How Risk Information Traditionally Is Handled—The Good, the Bad, and the Ugly

Regardless of whether or not one embraces and invokes the monetization process that is the heart of this treatise, there are some elements of the risk assessment and management processes that are ubiquitous. Just one of the universal components is the "risk register."

I put the term risk register in quotes because the term means different things to different people and groups. In this book, I attempt to take the high-level approach and loosely define the risk register to simply be a means of recording and organizing the risk-related data and information. It matters not in the following discourse whether the register is kept on paper or electronically, whether risk information is stored in a formal database or a spreadsheet, whether the storage device is "on line" or isolated on a particular hard drive, whether the register is generally available to all involved or is lorded over by a keeper or the like. It is simply my intent in this chapter to elucidate the critical elements of a typical risk register and to illuminate some of the issues that can elude the uninitiated and, ultimately, damage the entire risk effort.

Risk Registers and Their Many Implications

A risk register—as an entity—is a good thing. It is the uses to which registers are put that can be objectionable. In Chapters 6 and 7 of this book, the reader will see that risk registers play an essential role in the two primary routes to risk monetization. While it is true that risk registers will be the focus of at least portions of later chapters, such registers also serve as critical vehicles in processes that do not lead to monetization. The nonmonetization route is the crux of this chapter and, thus, a preliminary discussion of the fundamental elements of risk registers is warranted.

First, I suspect I should define the terms—that is, what I mean when I employ the term "risk register." With today's internet-based technologies, risk registers are commonly Web based—typically hosted on a company's intranet. This means that the register is a dynamic and multifunctional beast that all register-related individuals—or, perhaps only a select few—can access for the purpose of updating. On the opposite end of the sophistication spectrum, a risk register can be a document—on paper or in electronic form—in which risks and associated information and data are recorded.

The middle ground between these two extremes is most common—the dreaded (my opinion) spreadsheet. A stylized sheet appears in Figure 2.1. The illustration of a risk register in Figure 2.1 is in *no way* meant to be representative of a comprehensive register. Rather, it is included here to impart to the reader the typical and "bare minimum" items that constitute most registers.

As can be seen in Figure 2.1, the essential elements of a risk register are the following:

1. Project name
2. Risk identifier
3. Short risk description
4. Risk description
5. Probability of occurrence (POC)
6. Impact units
7. Range of impact
8. Translated impact—Common unit
9. Threat-mitigation or opportunity-capture action
10. Cost of threat-mitigation or opportunity-capture action
11. Mitigation- or capture-action owner
12. Mitigation- or capture-action implementer
13. Due date for action completion

Again, the register shown in Figure 2.1 is meant only to be illustrative. There are as many variations on this theme as there are risk registers in use. In the ensuing paragraphs, I will relate some of the more common variations on this general register outline.

In the risk register shown in Figure 2.1, "Probability" is represented by a single percentage. This solitary metric typically is gleaned by engaging in a conversation with a discipline expert in which the expert is asked: "How likely is it that this risk will occur?" or words to that effect. It is not uncommon, however, to represent the POC as a range. That is, it might be more effective—depending on the goal of the risk assessment and risk management plan—to express uncertainty about the POC. In such a case, a minimum, mean, and maximum percentage (or some variation of those three values) can be collected so that this probability can be represented in any stochastic analysis as a distribution.

Such an argument and representation can also be valid for risk register items such as costs. Note that in the register illustrated in Figure 2.1, the

Risk ID	Concise description	Risk description	Probability (%)	Impact units	Impact minimum	Impact peak	Impact maximum	Translated impact (NPV in $MM)	Mitigation/ capture action	Action cost	Action owner	Action implementer	Due date
COM 17	Government overthrow	Click here for full description	60	CAPEX in Millions $ (MM)	1	3	9	5	Early talks with opposition party	$0.5MM	Joe Smith	Jane Doe	1/1/20X

FIGURE 2.1
Stylized risk register in a spreadsheet.

"Action Cost" is represented as a single value. Reasoning similar to that outlined in the preceding paragraph can be applied. It is a matter of practical application and ultimate usage of the collected metrics that should dictate whether or not an expression of uncertainty is required to represent a given risk register item.

Although the entities represented in the risk register might be interpreted as relatively "transparent" and clear with respect to their meaning, I can tell you from experience that the label for each item is only a front for a more convolute real-world connotation. Therefore, I will in the following paragraphs address just some of the hidden context for essential risk register items.

Project Name

An entry such as "Project name" can seem almost arbitrary and innocuous. However, the moniker for the project should be an item not easily defined.

Creating a risk register is an exercise in listing and defining all of the threats and opportunities that relate to the project. In a gathering of project discipline experts at a risk-identification brainstorming session (more on this in Chapter 6), it is a given that each representative has in mind a different project—even though they would all swear that they had gathered to address the same project.

Just for example, a construction engineer at such a meeting likely is thinking about a project in which he or she will begin to build the physical infrastructure. Such construction is well down the road in the project's life cycle relative to other aspects. That construction engineer might be seated immediately adjacent to a negotiator who is imagining a project that begins with how the company will negotiate political and commercial terms with the host government—all things well in advance of "the cutting of steel" that dominates the engineer's thoughts. And so it is with all representatives at the meeting. Each person has a unique interpretation of just what is the project. For the sake of identifying and delineating risks, it is absolutely essential that there exists a common and consensus view of just what constitutes the project. That confluence of consciousness most certainly should be reflected in the expression of the project name.

I have found it essential to begin each risk-identification workshop (RIW)—that is attended by at least one representative from each critical discipline—with an address by the project manager or, perhaps, the commercial analyst. The purpose of the presentation is to clearly define exactly what constitutes the project. Such clarity in vision aids enormously in the identification and capture of a holistic set of risks and broadens the interpretation of the project for all attendees.

So, if the project is, just for example, to assess the viability and practicality of building a new manufacturing facility in a foreign country, a project name such as "New manufacturing plant" likely is inadequate. A title such as "Assess the full life cycle Net Present Value (NPV) of a widget-producing plant in Foreignland" might be more appropriate. The level of detail conveyed in the project title should be both comprehensive and practical.

Risk Identifier

Here I am skipping ahead a bit to Chapter 6, but sometimes such logic jumps just can't be avoided. You will see in Chapter 6 that at a RIW, attendees will ultimately be assigned to one of several predetermined high-level-category groups. One such subset might be the "Commercial/Competition" group.

After performing the subgroup tasks to be outlined in detail in Chapter 6, each risk identified by the group should be given a short "risk identifier." This is done for convenience sake and for easy reference to specific risks when discussing and ranking the risks.

For example, if the Commercial/Competition subgroup ends up with 25 risks in their high-level category, they might simply identify each risk in sequence such as the following:

Com-1

Com-2

Com-3

And so on. Similarly, an Operations high-level-category group might utilize risk identifiers such as the following:

Op-1

Op-2

Op-3

Short Risk Description

Because it is cumbersome in the risk register to display the full description of a risk—which sometimes can be quite lengthy—in the typical view of the register, it is often handy to have generated a few words that give a hint of the nature of the risk. The Risk Identifier described in the immediately preceding section should not attempt to convey such a description.

If, for example, the full description of a risk is

> If contract negotiations with the labor union fail to result in a satisfactory contract before the start-date for the project, then significant schedule delays will be realized resulting in increased CAPEX and decreased NPV for the project,

then the short risk description for this risk might be as follows:

> Contract negotiations

Such short descriptions can be handy when sorting through the sometimes huge volume of risks looking for all risks that pertain to a given subject.

Risk Description—How to Express a Risk

First, whether the reader agrees that the term "risk" should indicate both threats and opportunities or the reader adheres to a more outdated and traditional view that risk uniquely implies threats, it is undeniable that when considering parameters that will comprise a probabilistic analysis, both threats and opportunities have to be considered.

It all starts with succinctly and practically defining the problem.

Let's again—as in Chapter 1—consider the situation in which the problem is defined to be to forecast your net income for next year. To accomplish this, you will build a probabilistic stochastic model. When you make a list of parameters that will impact your future net income, you certainly have to include the threats or "downside" impacts—the things that have traditionally been considered risks. Items such as taxes, limited working hours, health care costs, and the like are obvious income-reducing considerations.

However, if you are to generate a credible forecast, such an estimate must include probabilistic opportunities or "upside" considerations. Variables such as the annual bonus, royalties from a book that is scheduled to be published, new investment returns, and other parameters have to be assessed and included. These things are not simply the "upside" extension of variables that would traditionally be considered threats or risks. Rather, these items are probabilistic events that will positively impact net income. So, whether or not you subscribe to the view that the term "risk" embraces opportunities as well as threats, there is no question that a credible estimate of your future net income has to include the probabilistic impacts of both "upside" and "downside" consequences.

So, a "risk" register would include both threats and opportunities and I will expound upon this much more in the next section. One of the primary reasons for expressing a risk is to plan what to do about it. That is, if we have declared that a probabilistic event is a threat, a salient response is to delineate actions that will mitigate the threat (I will address the meaning of "mitigation" much more in an upcoming section of this chapter). Similarly, if a probabilistic event has been identified as an opportunity, a critical element of addressing that opportunity is to define actions that will enhance the chance that it is captured.

Description of any threat-mitigation or opportunity-capture action must be preceded by a succinct expression of just what is the threat or opportunity. It is exceedingly difficult to define mitigation or capture actions for vaguely described events.

As you likely know, I have been a practitioner in the risk arena for a long time. During that tenure, I have been witness to the construction of countless risk registers in which threats and opportunities are anemically defined. For example, the schedule for a construction project might be adversely impacted by inclement weather. It is not uncommon to see such a threat expressed as "Bad weather." It is just as likely to see opportunities similarly stated. For this construction project, there might be a unique opportunity for early sale of the product if the factory is completed ahead of the currently accepted schedule. It is not uncommon to see such an opportunity expressed as "Early sales" or as some other equally cryptic moniker.

If it is part of the goal to outline threat-mitigation and/or opportunity-capture plans (and it is), then such descriptions are woefully inadequate. Best practice in the risk-description department is to attempt to relate each threat or opportunity in the form of Reason–Event–Impact.

For example, the "Bad weather" threat could be better described thus: If the monsoon season is of greater duration than the expected 3 months (Reason), shipment of construction elements overland will be curtailed and more expensive water-shipment methods will have to be employed (Event) causing construction delays, increased capital costs, and a lower NPV (Impact). The "Early sales" opportunity might be expressed as follows: If the plant is completed at least 6 months ahead of schedule (Reason), there is a one-time opportunity to sell product to a local utility company (Event) resulting in significant revenue and increased NPV (Impact). I find that there are exceedingly few threats or opportunities that cannot be expressed in this format. When the Reason–Event–Impact format is employed, it is a much simpler task to create plans to capture the opportunity or mitigate the threat.

Correlations between Threats and Opportunities

First, let's be clear—or, as clear as we can be—about what we mean by the terms "threat" and "opportunity." Again, I realize that Chapter 3 is the place for detailed definitions, but to address the merits and pitfalls of risk registers, I have to at least make rudimentary reference to these two terms.

Some opportunities are just "frowns turned upside down." For example, if a coin-flip "tails" represents a threat, we could put in our risk register that there is a 50% chance of this "downside" or "bad" thing happening. The "glass half full" guy, however, might look at the same situation and record in the risk register that there is a 50% probability that the opportunity associated with a "heads" might occur. Both items, of course, might be recorded which can lead to trouble.

In real life, I have witnessed this threat–opportunity duel play out on innumerable occasions. On large projects, it is not uncommon to have the RIW divided into several working groups. One group might focus on commercial and financial aspects, while other groups turn their attention to health and safety, operations, design and construction, and other major elements of the project.

Working independently, it is exceedingly common for different groups to realize the same risk but to express it somewhat differently. For example, the folks working on the commercial or financial aspects might identify and describe a risk thus:

> If the current political party wins the next election, then current existing hurdles in obtaining permits will stay in place, significantly increasing our construction time and costs.
>
> Clearly, this working group sees the political situation in the host country as a threat.

Conversely, those working on the design and construction considerations might identify and describe political risk thus:

> If the opposition party wins the upcoming election, currently existing barriers to obtaining permits might be reduced or eliminated which will considerably enhance our project schedule and reduce our construction costs.

It is standard practice to look for risks that are expressed more than once. An example might be the health and safety workshop group expressing a threat:

> If we are constrained by the host-country health and safety policies, we could experience a significant increase in injuries, thereby increasing our costs.

Looking at the same threat from the commercial or financial group's point of view, they might express a risk:

> If we can't modify the commercial terms of the contract with the host country which include and are dominated by health and safety regulations, then we could realize somewhat higher costs due to schedule delays from lost-time injuries.

Although somewhat differently stated, it is relatively clear that both groups are considering the same scenario. It is best practice to review all of the stated risks—and there could be hundreds of them—in an attempt to assure that the project group is not "double dipping" a risk. In the workshop, the groups will assign probabilities and consequences to the risks. In this case, the consequence likely will be expressed in dollars because the consequence in both cases is cost. If this double-capture of the same risk is not caught in the review of the risk register, then hopefully the additive consequence (double the costs) will catch someone's eye and the multiple accountings will be discovered.

If the double accounting is not discovered for two (or more) threats, or, for two (or more) opportunities, it is bad enough. This will, in this case, result in too-high costs being assigned to the risk. Although this is not what you want, it is not nearly as insidious as not catching the double (or more) expression of a risk as both a threat and an opportunity.

Consider the first example described in this section—that of an upcoming host-country election that could significantly impact permit-acquisition costs. One group—the one that viewed the election as a threat—will generate a consequence that is an increase to the project's base-case costs. The other group—the one that perceived the election as an opportunity—will assign a project-base-case cost reduction to the risk. If, just by happenstance, both groups independently assigned the same probability and the same magnitude cost and cost reduction to the risk (yeah, like that would happen, but allow me this bit of silliness), then the risk would be cancelled out. With regard to financial impact on the project, this is the same as not identifying the risk at all.

This is worse than double-dipping the consequence of a risk because at least when you double dip, you are accounting for the impact of the risk and you have some chance that the double-magnitude consequence will seem out of bounds to an astute observer and the double accounting will be discovered.

However, when a threat and opportunity cancel one another out, the impact of that risk is not seen. Seeing nothing is ever so much more difficult to do than seeing a double-magnitude impact. In addition, if the magnitude of the consequence of the threat is relatively large and would have a huge impact on the project, an equal-in-magnitude reduction of costs associated with the opportunity will still result in a near-zero impact for the risk. This is bad.

Correlation of threats and opportunities in a risk register is a long-standing and difficult-to-remedy foible. The best countermeasure is to have the risks in the register "regrouped" after the RIW by a group of people who did not serve in any of the workshop subgroups. For example, if the new independent cadre perceives that political considerations should be brought together from all of the risks described by the original workshop subgroups, then it is likely that the threat–opportunity conundrum will be discovered. Eternal vigilance is the only good advice regarding this issue.

Probability of Occurrence

This aspect of the risk register is more complex to address than the moniker might imply. POC is inextricably tied to other aspects of risk analysis.

I will first point out that probabilities are typically, but not always, expressed as a percent—that is—in parts per hundred. A statement such as "There's a 60% chance that we won't be granted access" is usual. If your risk analysis is headed toward monetization of the risks, as this book would imply is the proper route in most cases, then probabilities expressed as percents is likely what you want.

However, it should not come as a surprise that people in different disciplines will find all sorts of ways to express probability and to interpret the term itself. Representatives from the Health and Safety arena, just for example, have a tendency to consider "frequency" to be the same as probability. That is, they might count how many times a particular task has been performed and make note of how many of those instances resulted in an injury. So, it is not unusual to hear from them statements such as "15 of 37 activities resulted in injury." I know this sounds mundane, but I can't tell you the number of times I have seen people convert the "15 of 37" to 15%. I know, I know, it seems obvious, doesn't it, but you might be amazed at how many times this mistake is made—even by arithmetic-savvy folks. Even if you do make the proper conversion of "15 of 37" to percent, take the time to ensure that the frequency count expressed can properly be converted to percent.

For example, the "15 of 37" expression might be a subset of the percent probability you really are seeking. It might be that there were 37 chemical-related processes performed 15 of which resulted in injury. You might, however, actually be interested in the sum of chemical and mechanical tasks and related injuries. The bottom line is, make sure that the "counts" (frequencies) expressed can be rigorously translated into the probabilities in which you are truly interested. This sounds mundane, I know, but sometimes vast tables of "counts" (frequencies) can be presented and it is no trivial task to understand and convert such statistics into the percent probability you desire. In Chapter 3, I will try to bring home the difference between frequency and probability by relating a story of the win–loss records of two sports teams and the prediction of which team will win an upcoming match.

Another common presentation of probability—especially in the engineering realm—is what I would call the "proportion" expression. Again, I can't tell you the number of times I have been presented with, "There's a one in five chance that ..." At first blush, this might seem the same as the "frequency" or "count" methodology explained above, but usually it's not. Folks who use the "X in Y chance" expression typically—but not always—are not thinking about counts of things. Their brains simply naturally view probabilities as do gamblers, and some of them are really good at it, issuing statements such as "... and so, I think there's a 7 in 26 chance ..." to express probability (it hurts my brain to even think about "7 in 26" or how you would tumble to that in the first place). Nonetheless, there is a large cadre of bodies out there, and maybe you are one of them, whose brains naturally view probability in this way and, again, they are *not* thinking about a "count" of 26 and a "count" of 7. There's nothing wrong with this, but be sure that when converting this expression to percent—just like the "frequency/count" example above—that you consider the big picture and that your conversion actually results in the probability that you need.

By no means do I imply here that the two examples given above constitute an exhaustive set of ways to express probability. I will never cease to be amazed at the novel ways people find to approach probability and to express it. I present these examples as just that—examples of a universe of interpretations and expressions of probability.

A risk is an event. In a preceding section entitled, "Risk Description—How to Express a Risk," I describe how a risk is properly expressed using the

Reason–Event–Impact format. A person's perception of the probability associated with a risk is directly a function of how that risk is described.

For example, in the previous section, I used the example of someone expressing a risk as "bad weather." If a person were queried regarding the probability of having "bad weather," they might envision thunderstorms, windy days, and other inclement conditions. The probability of these types of weather conditions might be fairly high in the part of the world being considered. A probability of relatively great magnitude would likely result. However, if the risk was properly expressed as follows:

> If the monsoon season is of greater duration than the expected 3 months (Reason), shipment of construction elements overland will be curtailed and more expensive water-shipment methods will have to be employed (Event) causing construction delays, increased capital costs, and a lower NPV (Impact).

Then, the specificity of the risk expression is likely to generate a lower and more practical estimate of probability because that likelihood has to address specifically the monsoon season, a delay greater than 3 months, shipment of construction elements, impact on NPV, and other constraints imparted in the risk Reason–Event–Impact definition.

In all of the examples given above related to probability, I have addressed the issue of probability as though only deterministic estimates can be made. That is, I have used single values to represent uncertainty—after being converted to percent.

As you will see in Chapter 3, however, probability is one of the risk-related parameters about which we can be uncertain. If you have read any of my other books, you are aware that when addressing expressions of uncertainty, I favor the "minimum/peak/maximum" approach rather than the P10/mean/P90 or other similar approaches.

A coin flip is a probability example that is grossly overused, but it is ubiquitously employed because it is so clear an illustration of the principles. In the coin-flip scenario, the probability of getting a "heads" is, of course, 50%. Under normal circumstances using a fair coin, there should be no uncertainty regarding the probability of getting a "heads" on any given flip.

However, in the real world, probabilities associated with risks typically are not so apparent. For example, you might be attempting to estimate the probability that the host government will issue to you the critical permits necessary to begin construction before the monsoon season ensues. There might be all sorts of political intrigue involved in the issuance of the permits. Therefore, any estimate of the probability of having the permits issue in time must be uncertain.

In this case, I would urge the collector of data to glean from the domain expert an estimate of the minimum, peak, and maximum values associated with this probability. Again, the strict meaning of these terms will be presented in Chapter 3, but I wanted to address here something about the methodology used to collect the data.

People just love to tell you how good it will be. If we are talking about good things we will get, such as revenues, people supplying information for a project in which they have a personal stake will tend to want to express how "wonderful" the revenues are projected to be. Similarly, if the parameter in question is something that might be detrimental to the project—such as a cost—those with a stake in the project will tend to minimize these things. Conversely, those who wish revenues to be maximized for the good of the project will tend to downplay the magnitude of the minimum value as will cost estimators downplay the magnitude of the maximum value.

When collecting information regarding the uncertainty associated with a probability (or, almost anything else), one should try to avoid first getting an estimate of the "peak" or "most likely" value ("peak" here refers to the highest point in a frequency plot). If a conversation is first had regarding this peak value, then you can experience a phenomenon called "anchoring." Anchoring implies that once a people have convinced themselves of what the value is likely to be, then it is exceedingly difficult to get them to "think wide enough" about that parameter. That is, it is quite a vexing task to get them to imagine or express minimum or maximum values that reflect reality when they have first convinced themselves of what the actual value is going to be.

To avoid anchoring and to get realistic estimates of probability from discipline experts, I recommend the following process. When discussing a parameter for which low values will be "good" (as for a cost), then begin the discussion about the minimum value. However, make sure to record the reasons why the person believes the coefficient they express is a legitimate estimate of the minimum value. For example, they might say that their low construction cost estimate is valid because of the following:

- The weather is predicted to be mild during the construction phase.
- Governmental agencies are likely to produce on time the required permits.
- Fuel costs are forecast to be favorable.
- We are likely to be able to use nonunion labor.
- We are likely to be able to fabricate critical elements in-country rather than importing them.

Now, having gotten an estimate of the minimum construction costs based on the justifications listed above, discuss the maximum construction-cost value and use the same, and more, reasons to get a fair estimate of the maximum. For example, consider the impact if the weather is bad, if the government does not issue the permits on time, if fuel costs are significantly higher than forecast, if we have to use union labor, and if we have to import the critical construction elements. That is, use their low-cost justifications "against them" in order to generate a rational maximum value. Such techniques should be used in reverse (discuss the maximum value first and record justifications) when discussing a parameter for which the maximum value is "good"—such as revenue, production, and the like.

In either case, the minimum and maximum values should be established before any conversation endues about the peak or "most likely" value. In this way, anchoring is avoided and reasonably realistic estimates of minimum and maximum values result.

Impact Units

Unlike probabilities which usually end up being expressed as a percent—regardless of how they were initially presented—the units utilized to express impact or consequence can be all over the map, so to speak. This issue is so onerous that I can only present general advice here and hope that the reader can translate that generic advice into valid project-specific expressions of impacts. A good piece of advice, however, is to keep the end in mind. That is, before you start a risk analysis, be sure you have a crystal clear idea regarding the units in which the "answer" will be expressed.

First, I suppose I should make it clear just what I mean when I say "impact." Rightly or wrongly, it is my wont to use the terms "impact" and "consequence" interchangeably. A consequence is simply a thing that happens if the risk materializes. An example of a consequence resulting from a risk, that is an opportunity, might be early entry to a country in which a manufacturing facility is to be built. Early access will give your company a significant advantage over the competition. Impact resulting from a threat might be painfully higher manufacturing costs due to changing regulations.

In the case of the opportunity briefly described above, what would be your choice of impact units? If you were considering schedule risk, you might opt for time units such as months. However, if you chose such temporal units, it might be a bit short sighted.

In the first paragraph of this section, I recommended that when considering impact units, it is important to keep the end in mind. If your schedule risk analysis is or is going to be an integral part of your overall risk effort which will culminate with an expression of probabilistic NPV, then perhaps expressing the early access in months—or other time units—is fine. That is, if the movement of events in time will be translated into monetary impact by the equations of the risk model, then expression of impacts of threats and opportunities in time units within the schedule risk analysis will yield the desired result—impact on NPV.

If, however, the schedule risk is not an integral part of the overall risk analysis, then you might consider expressing the impact from the same access opportunity in, for example, dollars per day. That is, early access will gain the company exclusive access to in-country resources, which will result in significant savings and sales advantages over your competition. This type of translation of impact into money might be uncomfortable, but remember that regardless of whether you do the translation in this manner, or, you translate the output of the schedule risk analysis into NPV, the translation has to be done nonetheless. The point is, you need to have a broad view of how the entire risk analysis will be handled in order to make up-front credible decisions regarding units.

It is my way to tell self-deprecating stories—mainly because there are just so many of them to relate. Early in my risk career, I did not follow the advice related above when performing a risk assessment for a major construction effort in a foreign country. One of the primary parameters of the analysis was production. In my zeal, I collected the impact in units of tons. I spent many days overseas collecting such information from more than a dozen people who were geographically scattered—not even all in the same country.

I had the opportunity earlier to visit with the person who had generated the financial spreadsheet that would be the foundation of the economic analysis and that would serve as the vehicle into which my risk-related information would need to be fed. As it turned out, the construction of the spreadsheet was a bit unconventional—or at least that's what I thought—and the data I had collected needed to be fed into the spreadsheet in units of dollars per ton and not as raw tons. The tons I had calculated were the result of some relatively complex calculations based on data collected from the aforementioned dozen or so and geographically scattered individuals. Because of the cost units associated with the various components that went into the calculation of the tons, it was not simply a matter of taking the tons I had calculated and multiplying them by some number of dollars. It was no small task, and a relatively embarrassing one, to have to revisit issues again with the same people just so I could get the data into the required dollars per ton units. So, have the end in mind and have a good idea of how the data you collect will be used.

I will offer one more word of caution regarding units. The end result of most risk analyses—but not all—is some sort of monetary expression. NPV, internal rate of return, return on capital employed, and other metrics are common. Most of these calculations involve discounted cash flows over time. It is not within the purview of this book to describe the various financial metrics, but given that most of them involve a time-series of cash flows, it would be remiss of me to not offer a caution here with regard to impact units.

You might be considering the threat of a strike at the preferred compressor manufacturer. If the labor strike materializes, you will have to purchase and employ a cheaper and inferior compressor. The inferior compressor will be delivered later to the project because it would be ordered only upon learning that the desired, and already ordered, compressor will not be available.

In such a scenario, the first thing you need to guard against is the temptation on the part of risk-register builders to allow entities such as "Compressor

Manufacturer" to serve as either the units of impact or as the impact itself. For example, the risk might be described thus: If Manufacturer A has a labor strike, then we will have to purchase inferior compressors from Manufacturer B, which will lessen our ability to move product effectively which will, in turn, lower revenues and increase operating costs.

In this case, it can be tempting to allow "Manufacturer A" or "Manufacturer B" to be valid entries in the risk register for either the impact units, or, the impact itself. If we have in mind, for example, NPV as the ultimate expression of the impact of integrated risks, then such a textual entry in the register will preclude use of this parameter in any risk-based equations except in an "if, then" situation (If we get Manufacturer A, then ...).

Furthermore, if the purchase of the compressor is a one-time cost—paid for in a single installment—then when you are collecting the data for the compressor purchase price, be sure to have a discussion regarding *when* the cost of the compressor will impact the discounted cash flows. The time value of money is important, and costs occurring later in the time series can have significantly different impacts than those costs occurring earlier in the series. In addition, if the cost of the item is to be dealt with over time, be sure, when you are discussing impact and units with the domain expert, to collect information relating to how many time units will be impacted, specifically which time units will be impacted (costs might not start impacting the project until some time "down the road"), and what proportion of the costs will be allocated to each impacted time. The point here is that you must select impact units that can be spread over time and that discussions of how consequences will be allocated have to be an integral part of the impact discussion and of the decision about units.

Range of Impact

Unlike the sections above, this one will be relatively curt. In the preceding Impact Units diatribe, I outlined how you should keep in mind the end result when selecting impact units. The same advice is valid for the impact itself. My tons vs. dollars-per-ton example is illustrative as is the suggestion not to allow entries such as "Manufacturer A" as an impact when risk integration and/or monetization is the aim.

As with the collection of a range of values for impact units, you should beware of anchoring and should begin your query of minimum and maximum values based on whether the impact is an opportunity or a threat. See section "Impact Units" for this discussion.

A final bit of guidance I will offer with respect to impact is that when integrating the impacts of risks, guard against unnatural, or impossible correlations—especially when ranges are involved. When deterministic values are used to represent impacts, it is relatively simple to avoid unrealistic combinations of impact values. For example, costs and revenues might be inextricably linked in the financial equations. That is, as costs go up, revenues will necessarily decline. If a simple spreadsheet is being utilized, then it is fairly elementary to fill in the cost cell with a value that makes sense relative to the value used in the revenue cell. However, when ranges are used, this is a more vexing endeavor.

You might have represented the costs with a range from very low to reasonably high costs. In a Monte Carlo process, this range would be represented by a distribution. On any given iteration of the Monte Carlo process, any cost value might be used to fill in the cost cell. Similarly, you might have indicated a low-to-high range for revenue. Like costs, on each of the Monte Carlo iteration, the revenue distribution will be randomly sampled to get a value to put in the spreadsheet revenue cell.

Clearly, if no sampling constraints are implemented, unrealistic combinations of costs and revenues will result (low costs with low revenues, high costs with high revenues, etc.). When using ranges to represent impacts, be sure to use whatever tools are available to constrain the selection of values for related parameters. Some software packages offer correlation tools. In other packages, you might have to write computer code to control sampling. Whatever you have to do, be aware that as soon as you move from the realm in which impacts are deterministically represented into the world of ranges, additional steps have to be taken to attempt to ensure that realistic combinations of coefficients are used in the analysis.

Translated Impact—Common Unit

In the section "Being Able to Translate that Expression of Risk into Something that Can Impact Value" of Chapter 1, I present a brief explanation of why it is important to translate the many expressions of risk into a common metric to be used for risk ranking. In the "Translation of Risks" section of Chapter 5 and in the "Create Translation Table" and "Use Translation Table to Convert to Common Measure" sections of Chapter 6, I present methods that might be used to translate disparate risk units into a common metric.

The focus of this chapter is mainly to identify and define the elements of a typical risk register. I address this important aspect of the risk-monetization process in other sections of this book. Therefore, I will here only define the "Translated Impact" register item.

Threats and opportunities will be viewed and expressed in myriad ways by the broad spectrum of project-pertinent disciplines. A major aim of creating a risk register is that of collecting data, so the risks can be relatively ranked. It is difficult to rank, for example, a risk that has an impact expressed in "tons" relative to another risk that has impact presented as "months of delay." If ranking is to be accomplished, then the expressions of risk impact must be translated into a common metric. The coefficient of that common metric (NPV in the example shown in Figure 2.1) for each risk should be recorded in the risk register. The aforementioned sections of Chapters 1, 5, and 6 address how such critical translation is enacted.

Threat-Mitigation or Opportunity-Capture Action

Threat mitigation and opportunity capture are subjects addressed briefly in Chapter 1 (Can't Afford to Implement a Risk Process) and will be delineated in detail in Chapter 3. In the aforementioned section of Chapter 1, I laconically presented rationales ubiquitously used to resist implementation of a risk process. Included were arguments relating to a lacking cost–benefit ratio, resources necessary to upset the organizational status quo, and the dearth of lines of sight between the risk process and its touted benefits. In Chapter 3, I will in detail address the definitions of critical terms including mitigation. In spite of the fact that counsel on this topic is recounted in other sections of this book, some of the more mundane characteristics of threat mitigation and opportunity capture must be presented here because these subjects are common elements of most risk registers.

I will start this discourse with a kind of story. Consider car companies and their attempt to make car travel safer. Management at these companies are cognizant of the fact that the probability of having a car accident is rising rapidly. Crumbling infrastructure, the "density" of cars on the roads, more young and old drivers behind the wheel, and other realities contribute to the sharply upward trend in the likelihood of being involved in motor incident. The car companies know that they can't do much to lower the probability of car accidents, so they focus on impact.

When I started driving, there were only lap belts in cars. Solid steel steering columns, steel dashboards, jagged hood ornaments, and other death-or-injury-inviting attributes were common in cars. Today, unibody construction, crumple zones to absorb energy, collapsible steering columns, padded dashboards, head rests, lap and shoulder belts, front and side airbags, ramps that direct an engine down to the road rather than into your lap, stability control, antilock brakes, and myriad other safety-related characteristics are typical or mandatory in cars today. The point is, car companies focus on impact mitigation because they know they can't do much to positively affect probability.

Now consider airplane manufacturers. I started flying for business in the late 1970s. When I entered the passenger cabin of the ubiquitous Boeing 727, the "safety" equipment included door-mounted inflatable ramps that could double as rafts, drop-down oxygen masks, floor lighting that lead to exit doors, and a lap belt. When I today walk into the cabin of a state-of-the-art aircraft, the "safety" equipment includes door-mounted inflatable ramps that can double as rafts, drop-down oxygen masks, floor lighting that leads to exit doors, and a lap belt. Hmmm, vaguely familiar stuff. Less flammable seat and other materials are required, "16-G" seats are used (they are less likely to break away during a crash) and defibrillators are now carried, but airplane manufacturers know that if they fly the aluminum (or composite material) tube into a mountain at over 500 miles per hour, there is no amount of safety stuff inside the tube that is going to make much difference with regard to customer survivability.

So, the airline industry in general focuses less on impact mitigation and more on reducing the probability of having an airplane-related mishap. Collision-avoidance radars, advanced autopilot systems, reams of new rules (not flying in downdrafts, for example), new spacing rules, and other accident-probability-reducing advancements have made airline travel the safety-related benchmark mode of travel. The point is, the airplane manufacturers and the airline industry in general focus less on impact mitigation and more on mishap-probability reduction—just the opposite of car companies.

You're probably saying to yourself: "Well, Glenn, that's a swell story, but what's it got to do with risk registers in this section of the book?" Believe it or not, it is relevant.

For the sake of brevity, I'll make an example here of only threat mitigation. However, be aware that every point made in the following text applies equally to opportunity capture.

As will be addressed in detail in Chapter 3, two attributes associated with every risk—threat or opportunity—are probability and impact. If you grasped the gist of the car/airplane story above, you understand that, with regard to a threat, one can take steps to mitigate the probability, the impact, or both. It absolutely is essential that when considering mitigation actions, you consciously decide and express in writing which of these attributes will be affected by the proposed mitigation action. It matters.

I will here revert to my, perhaps overused, example of the risks associated with building a manufacturing facility in a foreign land. One of the risks identified is that of the government holding up issuance of required permits. In the risk register, this risk is expressed in the recommended Reason–Event–Impact format thus:

> If, for political and economic reasons, the host government puts unnecessary hurdles in our permit-obtaining path, the optimal window of construction for this year might be missed resulting in significant increase in logistical and construction costs and increased opportunity for our competition.

A first concept I would ask you to understand is that, for this risk, you can decide to propose mitigation actions that address the probability of the threat materializing, the consequence of the threat, or both. It is good practice to specifically discuss possible mitigation actions for both probability and impact.

In this case, we might consider a mitigation action to reduce the probability of materialization of this threat. One action to be taken could be to send an emissary to the host country on a "listening campaign." The point of this effort would be to meet with critical host-country officials to learn about their concerns and to demonstrate that we care about their issues. Preemptive suggestions and offers that address their concerns might lessen the likelihood that host-government officials will erect the procedural hurdles that could delay our company's progress.

At the same time, you might consider implementation of mitigation actions that will lessen the impact of the threat should it materialize. For example, you might consider this year importing and staging construction equipment and materials that would be required in initial and subsequent construction phases. Such relatively expensive but possibly prescient staging of materials will allow immediate construction next year (rather than starting the import of material next year) should the threat materialize.

So, the bottom line is that when recording mitigation (or, opportunity-capture) actions in the risk register, every threat mitigation action should begin with the following words:

"To reduce the probability ..."
or
"To lessen the impact ..."

Or other such words that specifically identify the action as one that aims at probability or impact. You should always consider actions that might address both probability and consequence and should identify those proposed actions as addressing one or the other attribute.

Cost of Threat-Mitigation or Opportunity-Capture Action

"There's no such thing as a free lunch" is an adage that aptly applies here. It is an exceedingly rare event when a threat-mitigation or opportunity-capture action does not bring with it a cost (however, we hope that cost is less than the cost we will experience if we don't take the opportunity-capture or threat-mitigation step).

Just like impact units (see the section on impact units above), the costs associated with a mitigation or capture action need to be expressed in units that are relevant to the answer you are seeking. For example, if the ultimate expression of the risk analysis is going to be NPV, then you should not allow mitigation or capture costs to be expressed as "slower progress" or the like. In any worthwhile risk-monetization process, these costs will be integrated with all other information to arrive at the answer parameter. So, if your aim is NPV, then these costs should be expressed in units that can directly influence NPV. If you have nonfinancial goals for your risk assessment—such as time—then the mitigation- or capture-action costs should be expressed in units that can directly impact schedule events (days of delay, for example).

Again, just like the advice given in the impact units section above, the influence of mitigation or capture costs might have to be spread over time. That is, regardless of whether the answer parameter is time based or financial, it is not uncommon to encounter temporal subsets, or a time series, in the analytical process. For example, the cost of an opportunity capture action might be deemed to be $1,000. However, the cost will not be encountered until, say, period (year, month, etc.) 4 of our *N*-period analysis. The cost will also not be entirely attributed to period 4, but will be spread over period 4 and the next two periods. So, when collecting cost data for this action, be sure to gather from the discipline expert the proportions of the cost that will be allocated to each time period. If this is done in percent, then perhaps 50% of the $1,000 will impact period 4, 35% period 5, and 15% period 6. If you are not thinking ahead with regard to mitigation or capture actions, you will likely have to reinterview discipline experts to obtain the necessary data. There is only so much patience people have for inefficiency.

Yet another cautionary note associated to mitigation or capture costs is related to their distinction from the impact of the risk. I can't think of a single good reason why the mitigation or capture cost should, in the risk register, be combined with the risk-impact cost.

It should be abundantly clear as to why impact costs and mitigation or capture costs should not be combined in the risk register, but I have been witness to innumerable such combinations. It's a bad idea, and here are just a few reasons why.

Let's say that we have an opportunity that could yield a probability-weighted revenue increase of $1,000,000. If the probability-weighed costs associated with actions that would increase the likelihood of capturing the opportunity were $999,999 and in the risk register we combined the impact cost and the opportunity-capture cost, we end up with an opportunity impact of $1. In the risk register, we often rank threats and opportunities by their probability-weighted impacts (try to "bubble" the "biggest" risks to the top of the pile). This opportunity—which could yield a $1,000,000 benefit, would not rank very high when being judged by its $1 integrated benefit. This is a mistake.

Now, I know what you are thinking—why not rank risks by their integrated (impact cost minus the mitigation or capture cost) cost? The answer is simple. You always want to rank risks based on their unmitigated impact costs because that is what is going to be the impact if the risk materializes. Those significant impacts should not be masked by offsetting capture or mitigation costs. Project planners need to be aware of the "big" risks that are out there regardless of the cost to mitigate or capture them. In addition, when a significant risk is identified and "bubbled to the top" of the risk list, it is often the case that more conversation ensues regarding what might be done about the risk. This new conversation often revamps the previous list of mitigation or capture steps that might be taken.

Now, having made a case for ranking risks based on unmitigated costs, let me unravel that argument. The point of this book is to convince readers that monetization of risks is an essential process and to convey to those same readers the concepts and methods to accomplish the monetization task. In my small mind, monetizing risks relates to value and not necessarily costs.

For example, it might be true that spending $1,000 to mitigate risks emanating from one specific discipline has greater value than spending $5 to offset threats originating in a disparate arena. The point is, it is not necessarily about the absolute costs, but is about the impact those costs will have on the value of the project. In later sections of this book, examples will be made of this principle. So, when ranking risks in a risk register, it is best practice to employ some mechanism that will give you even a ballpark estimate of what impact a cost (expressed in money or time units, or whatever) will have on value (see "Translated Impact—Common Unit" section of this chapter). The means of making such estimates will be the subject of subsequent sections of this book.

Mitigation- or Capture-Action Owner

Very uncharacteristically, I don't have too much to say about this ubiquitous risk register component. In small enterprises, it is more likely that the risk mitigation- or capture-action owner and the risk mitigation-capture implementer will be one in the same person. That is, it might be determined at the RIW that Joe is responsible for seeing that the potentially defective critical valve is repaired or replaced. Then, not long after the meeting, it is Joe who jumps in his pickup, drives to the site of the valve, breaks out his rusty/trusty tools, and sets about fixing the valve.

On most project teams, it is a rare instance that the action owner is the same person as the action implementer. On a medium-to-large project team, it is customary that someone at the RIW, or, someone of rank who is designated by a workshop attendee, will be the action owner. When I say "of rank," I refer to someone who has the organizational clout to cause the action to be taken.

For example, the Corporate Director of Engineering, Jane, might attend the RIW serving as the mechanical engineering domain expert for aspects of a major plant-expansion project. At that workshop, a plethora or mechanical engineering-related risks are captured, including one that relates to potentially faulty valves that already exist in the plant that is being expanded.

As the Corporate Director of Engineering, Jane's purview includes a spectrum of responsibilities much more broad than those relating to this specific, but most important, project. Jane's attention will be directed in many ways in the near future, so at the risk workshop, Jane nominates Jack to be the risk-action owner with regard to the potentially faulty-valve issue. Jack is the Chief Engineer at the plant that is to be expanded. It is Jane's responsibility to alert Jack to the risk register, to the valve threat, and to his task. It is Jack's charge to make sure that the potentially faulty valves are fixed or replaced.

Jack's tasks do not end there, however. It is up to Jack to assign someone to actually fix the valves, to make sure the action is taken, to update the risk register with the appropriate remarks regarding how the problem was addressed, and to inform Jane that the issue has been resolved. Even for a single risk, this can be a daunting list of obligations. If Jack's already full plate has added to it the accountabilities associated with multiple risks, it is likely that Jack will be somewhat to completely remiss in being 100% compliant with the risk-related tasks.

Seeing this coming is Jane's responsibility. If Jane took the time to attend the RIW, then she likely deems the project to be salient. Hence, it falls to her to make certain that Jack is not overwhelmed by the added risk burden and to see to it that Jack has the help he will need to achieve 100% compliance. Each risk listed in the risk register necessarily implies obligations on the part of multiple employees (or contractors) and to not have a plan for practical distribution of responsibilities is to have a plan for risk-process failure.

Thus far, we have seen how Jack and Jane play a role in the scheme of things and we know that Jack, as Chief Engineer at the plant, is not likely to try to personally fix the valves. For this task, Jack will assign a risk-action implementer. This is the subject of the following section.

Mitigation- or Capture-Action Implementer

In the preceding paragraphs, I presented roles for management types, but at some point in the process, we have to get to the pointy end of the spear, so to speak. It is Jack's responsibility to select Joe, the Master Pipefitter, to actually get his hands dirty in fixing or replacing the critical valves.

Joe needs to be made aware, by Jack, of just exactly what is the problem—as stated succinctly in Reason–Event–Impact format in the risk register—and has to be made aware of the type of information Jack will need to satisfy the threat-mitigation requirements. If Joe is not apprised of these things before he embarks on the maintenance mission, he likely will not report back to Jack the critical elements Jack needs to sate the gatekeepers of the risk register.

One of the salient points of this section is that everyone up and down the risk chain needs to be made cognizant of the specific description of the risk in the risk register, of the action to be taken to mitigate the threat or capture the opportunity, and of the information that needs to be recorded—post action—in the risk register so that the risk can be reduced in severity or eliminated. Failure to inform all parties will lead to failure of the risk process.

Due Date for Action Completion

This seemingly simple item is actually one of those vexing commitments that harbors the capacity to torpedo even the most meticulously planned risk process. Contentiousness almost universally stems not so much from the action-completion date itself, but rather from which individuals set that date.

I have repeatedly pointed out in my ramblings in four previous books and in this treatise that risk assessment or management processes—and the successful implementation of those processes—is less about technical prowess and is much more about the adroit handling of people, organizations, and cultures. As described in the previous section of this chapter, it is essential that all affected parties—up and down the chain of command—be brought into the risk assessment or management or monetization process. And so, it is for the setting of the action-completion date.

Again, as described in the preceding sections, the identification and description of the risk, the estimates of related probabilities and impacts, and the delineation of opportunity-capture or threat-mitigation plans can involve individuals at various levels in the organizational structure. If the date for action completion is set at the RIW early on, then it is not uncommon that the date will not much reflect reality.

In the preceding sections, I made example of Jane, the Corporate Director of Engineering, Jack, the Chief Engineer, and Joe, the Master Pipefitter all playing roles in the risk process. If the mitigation action-completion date for the threat is set by Jane at the initial RIW, it likely will have less credibility than a date agreed upon by all three parties in this example.

Jane might set a date of February 10 of the current year for action completion. However, Jack could be aware of a partial plant shutdown for general maintenance that would make meeting the February date impractical if not impossible. His estimate might have been April of the current year. Joe would in turn not see the April date as a practical target because only he is aware of the vacation schedule for critical individuals and would not agree to a date any earlier than May of this year.

Now, admittedly this is a contrived example, but it is exemplary. If, as the result of the initial RIW, Joe has foisted upon him a risk register that contains unattainable goals, his confidence in the entire process is at least diminished and at worst shattered. The people that Joe directs are the "pointy end of the spear" so to speak, and are those upon whom rest the success of the entire enterprise. If these critical personnel believe the entire process is a joke because of, what is obvious to them, completely unrealistic targets that they think any fool would know they can't hit, then the risk process is mainly doomed.

So, the setting of action-completion dates might at first glance appear mundane. However, the dates set directly impact the individuals who will actually carry out the work necessary to capture the opportunity or to mitigate the threat. Those personnel have to believe in the dates and need to understand the rationale behind the dates if confidence in the entire risk process is to be maintained throughout the ranks. This is a very important element of the entire risk process.

Gatekeeper or Free-for-All?

Preceding sections have addressed specific elements or components of a risk register—Impact Units, Project Name, and the like. This final section dealing specifically with the characteristics of the risk register itself—rather than with how it might be employed—will address the technologies utilized to build risk registers and the associated implications.

I find that attempting to broach this topic is a bit like telling the story of the chicken and the egg, or, like trying to decide where to hop on the Merry-go-

Round. The attendant logic train is a loop rather than a linear concept, so it matters not much where the story begins.

One of the exceedingly practical decisions to be taken early in the planning of the risk register is just who will have access to it. That is, who will have authority, and more to the point, the responsibility to receive, actively seek, and record raw register data and updates? The aforementioned "loop" aspect of risk register has to do mainly with this conundrum and how it relates to what technologies are used to build the risk register.

If a register is created on the back of a brown paper bag or, only slightly more advanced, in a simple spreadsheet, then it is most likely that one person—typically the risk-process proponent—who will receive and record data in the register. This can be a daunting task for a multitude of reasons.

First, if there is one person or a small number of people who are sanctioned with register-changing authority, it could also be true that the anointed one also is responsible for monitoring the register and "bird dogging" the organization. For example, as action-completion dates become the current date, it might fall to the risk proponent to chase those responsible for action plans to comply with the date and to deliver the required information regarding the action completion. If the risk proponent (risk register "keeper") does not have authority to enact some sort of reward or penalty related to timely delivery of required information, then the impetuous of responsible parties to deliver such information can be near nil.

Especially in a "check-the-box-mentality" organization, it is common for that organization to view the creation of the risk register as a task to be completed. When it is done, it is not atypical for the register to be put on a shelf and not regularly referred to—much less envisioned as the driving force behind how business will be done. If this attitude permeates the organization, then the risk proponent—responsible for seeing that the obligations recorded in the register are met—is in for a rough ride. It is up to top management to imbue the risk proponent with the requisite authority to make the risk register an accepted (rather than excepted) centerpiece of daily operations.

A risk register is a dynamic entity. That is, after it is initially created and filled with data, it can't be seen as "done" if the risk process is to yield maximum positive impact on the project.

When first created and populated, the risk register represents a snapshot of the situation at the time of register creation. As time goes by, some risks will be captured or mitigated and should no longer be active elements of the register. More importantly, however, new risks—both threats and opportunities—will be revealed as the project progresses. The question is; just how will these new risks be identified, described, and captured in the risk register? This is a tough nut.

In the scenario of a single person or small group of people acting as the risk-register gatekeeper (the individual or individuals with authority to change risk-register data), there are two classic methodologies used for to identify, describe, and record new risks through time. One method is to regularly hold "mini" RIWs championed by the risk proponent. During these workshops, the list of existing risks is reviewed, progress discussed, and new risks captured and recorded. This process has the advantage of taking the entire burden of identifying register updates exclusively off of the risk proponent. However, such regular meetings, unless attended by top management, can be, as time goes on, as a bother and such meetings will be less and less attended by critical personnel.

Another classical approach used to update the register with new risks—when the register-gatekeeper process is enacted—is to have the risk proponent regularly "make the rounds" to responsible parties in the various disciplines to review already recorded risks and to capture any new risks. Of course, in this scenario, the risk proponent feels solely responsible for capturing new risks and for retiring old ones and can be seen as persona non grata by the organization ("Quick, hide—here comes the risk proponent, Sam!"). Again, if top management is not in favor of and is not openly supportive of the actions of the risk proponent, then that person's job is going to become increasingly irrelevant. I briefly address the issue of "keeping it fresh" again in the "The Importance of Keeping It Fresh" section of Chapter 5.

Well, thus far, I have described situations in which relatively simple technologies—spreadsheets or less sophisticated means—are used to create the risk register. These technologies were linked to a single person or a small number of people who can be sanctioned to build, populate, and update the register. However, modern technology—especially the internet—offers risk-register options that would have been unthinkable not long ago.

Today, it is common for risk registers to be Web based. That is, a register might be housed in a secure place on the company's intranet, or, on the internet. When this approach is taken, the risk register itself is seldom a simple (or even sophisticated) spreadsheet. Myriad software products exist that facilitate the construction of elaborate and technically advanced risk registers linked to powerful data-manipulation and storage packages.

In the relatively simple-spreadsheet scenario, it is likely that such a register exists on one person's PC hard drive and, therefore, it is a sticky task to allow multiple people to practically access to the register (not that there are not ways to do this). However, when a Web-based risk register is used, everyone who is privy to the required passwords and so on can simultaneously "access" (that is, operate on an image of) the risk register. This, then, begs the questions (among many others):

Who can access the risk register?

Who can update parts (and which parts) of the risk register?

Who can add new risks to the risk register?

And so on. These are critical issues. For example, if you are in the Commercial department of a corporation and are responsible for updating the commercially related risks in a Web-based register, you might not be fond of the idea that someone from the Law department could access your commercial risks and edit them. So, even in a Web-based scenario, it is smart to limit the number of people who can be responsible for accessing the risk register and to further limit the sections of the register that each of those people can access.

It's not as simple as that, however. What you don't want is people being able to access the parts of the register for which they are responsible only to find that risks for which they have some culpability get "edited away" without those risks having been fully addressed. This needs to be guarded against. The best way to prevent spurious entries is to make it the risk proponent's job to review all edits to the register. An e-mail file of all edits can be sent to the risk proponent on a daily basis so that he or she is aware of all changes.

With a Web-based register, it can also fall to individual discipline experts to monitor the project from the vantage point of their arena and be on the watch for new risks that arise. It is their responsibility, therefore, to update the register with those new risks. Again, the addition of any new risk should trigger an e-mail to the risk proponent so that he or she can monitor the new risks and verify that all pertinent information has been entered in the format required. An example of a battle for control of the risk register is described in Chapter 6.

Advent of Web-based risk registers brings to the fore an entire spectrum of interesting and sometimes not-so-easy-to-address issues—far too many to address here. It is my job, however, to make the reader aware of the fact that the leap from relatively simple spreadsheets to Web-based applications brings with it a new universe of capabilities and problems, and some of those organizational, cultural, and workflow conundrums are not easily resolved. Be prepared!

Suggested Readings

The list below includes material that the reader might find helpful and relevant to the subject matter discussed in this section of the book.

Barton, Thomas L., William G. Shenkir, and Paul L. Walker. *Making Enterprise Risk Management Pay Off: How Leading Companies Implement Risk Management.* Upper Saddle River, NJ: Prentice Hall, 2002.

Campbell, John M. *Analyzing and Managing Risky Investments.* Norman, OK: John M. Campbell, 2001.

McNamee, David. *Business Risk Assessment.* Altamonte Springs, FL: Institute of Internal Auditors, 1998.

3

The Fundamentals

In this chapter, I will relate the fundamental precepts of risk assessment and risk management—especially as they relate to the process of risk monetization. I would suspect that the reader expects this to be all about terms, definitions, and the like. However, many years in the applied-risk business has taught me some "basics" that you won't find in the typical text on risk. The immediately following sections are examples of such lessons from the school of hard knocks.

Decide for Whom You Are (Really) Working

It should be easy. Someone gives you a call, sends you an e-mail, or engages you at a meeting. They propose that you help them with a particular risk problem at their place of business, for example. After the typical negotiations regarding scope, price, payment schedules, and others, you show up at the designated place and dig in. Well, wouldn't it be just swell if it was all that easy. It's usually not.

Most humans—and I've actually known a few who seemed not to be—are cursed with a conscience and a learned or built-in set of ethics. Darn it all if we don't find that we mainly want to do the right thing. Just before I retired from my primary career, I found myself in a situation similar to the one I use as an example below. I have completely changed the nature of the business in the scenario related here, but the challenges and dilemmas are coincident with my experience. So, here's the story.

Mineitall is a mining company with specific expertise in mining, extracting, and converting to a commercial liquid the rarely found mineral Leaverite. Leaverite is found in relatively few places in the world where it can be economically extracted. A liquid mainly composed of the dissolved mineral, however, has applications in many industries and is highly sought. Concentrations of Leaverite in even the richest ores is in the order of 1–3% by weight (that is, 97–99% of the rocks in which Leaverite is found is heavy waste material).

It has long been known that a small land-locked foreign country—Foreignland—harbors most of the world's supply of Leaverite. However, a combination of low concentration of Leaverite in the ore, political unrest in the country, poor access to the mining areas, low commodity prices, and other global-economic realities have combined to discourage foreign investment in the Leaverite-rich nation. In Foreignland, however, there are a dozen or more small, inefficient, indigenous Leaverite mining operations that extract the mineral and convert it—*in situ*—to the commercially desirable liquid phase. There also exist dozens more potential mining sites that have yet to be exploited. The quality of the ore, accessibility, and other factors for the undeveloped sites vary widely.

A recent uptick in "green" technologies has caused significant increase in demand for the "liquefied" Leaverite (Leaverite is just the main component of the commercial liquid). Mineitall management had predicted the increased interest in Leaverite products and had more than a year ago negotiated exclusive mining rights to most of the desirable Foreignland acreage. Because Mineitall management realized that its Leaverite-related fortunes would lie in both acquisition of already existing in-country mining assets and with developing new mines, they deemed it reasonable to create a new department which they anointed "Business Development" (BD). The former leaders of the Mining Department and of the Acquisitions and Divestitures Department were to be coleaders of BD.

Over the past year, BD has successfully acquired nine of the already existing Leaverite mining operations and had created a single in-country business entity to oversee the now combined assets. In the past, each of the small production plants would ship the Leaverite liquid to market in rickety old World War II vintage trucks over roads that could only be termed "roads" because they were not jungle. Because Foreignland is landlocked and the Leaverite liquid needs to get to a sea port, the trucks had to make their way across neighboring foreign territory. Hijackings, "tolls" and many other practical problems regularly impacted the shipments. For Mineitall, this could not be the preferred method of getting the liquid to market.

Joe, Sally, and Mark all work for Knowitall Consulting (KC). KC has special expertise in the mining arena. Joe joined KC after retiring from another firm at which he was responsible for dealing with holistic risk assessment and management (RA/M) in major projects. Therefore, his view of risk includes the logistical, legal, commercial, financial, security, construction, engineering, research and development, political, and other aspects of a project. Sally and Mark are classically trained in mining with many years of academic and practical experience between them. Mineitall has asked that KC send some representatives to Foreignland to meet with Mineitall personnel.

Thus far, the Mineitall track record is 100%. That is, they have successfully acquired every already existing mining operation they have sought. Better yet, there are at least half a dozen more already existing operations ripe for the picking and more than 20 potential mining sites that have yet to be exploited. The BD guys at Mineitall, however, need some help in prioritizing their next few acquisitions—thus, the call to KC.

When Joe, Sally, and Mark arrived at Mineitall headquarters in Foreign-land, they had arranged meetings exclusively with BD management. BD management, of course, get rewarded for getting just as many existing and potential producing sites in the Mineitall portfolio as they possibly can—more is better. In their first meeting, Sally and Mark were fully engaged with BD management—reviewing geological maps, assessing assay values, considering locations, and so on for each of the potential new acquisitions. Joe, however, sat quietly.

Joe had been mainly staring at a world map on the wall of the conference room. He had also been tapping away on his computer in an attempt to learn more about the politics, religions, economies and the like of all of Foreignland's neighboring countries. BD management and Sally and Mark had for half a day been hot and heavy into assessing the potential acquisitions when they finally noticed that Joe had not contributed one word to the conversation. When finally asked why he was silent, Joe indicated that if he spoke, they would not like what he had to say. However, they insisted.

Joe fully realized to whom he was speaking. The audience was comprised of people who got rewarded for generating and capturing new Leverite opportunities—either already existing operations or the acquisition of acreage with Leverite potential. His analysis of the overall situation would not be well received.

Joe began his diatribe with the bottom line. He proposed that Mineitall should follow one of two paths. These paths were the following:

1. Mineitall should cease acquiring already existing Leaverite operations and should stop investigation of yet-to-be-developed acreage until the company can determine how it will:

   • Practically and economically link the already acquired mining operations with respect to ore transport and processing.

   • Economically and in an environmentally responsible way deal with the waste from Leverite processing.

   • Select a route to a sea port for a pipeline. This will involve negotiations not only with numerous foreign governments that control counties between Foreignland and the sea, but with warring factions within some of those countries.

   • Determine how pipelines between Foreignland and the sea will be secured.

   • Calculate whether all of this could be done at a profit.

2. If Mineitall does not have the resources to economically develop the entire concession, then it should cease acquiring already existing Leaverite operations and should stop investigation of yet-to-be-developed acreage and seek a buyer (or partner) for the entire

concession. Mineitall already has a 100% track record of successful acquisitions and has an already existing portfolio of potential
Leaverite mining sites. Attempting to acquire more existing operations or to develop more potential sites would be expensive and
might besmirch the currentlyexisting perfect record. The optimal
time to seek to sell the entire concession was the current time. Any
money spent in further acquisition or development would likely not
be recovered by a higher selling price associated with a less-than-
perfect track record.

Of course, when Joe related this advice, the others in the room looked at
him like he had two heads. Clearly, he and the other Knowitall consultants
were called to Mineitall to meet with the BD representatives—those who get
rewarded for acquiring existing operations and recommending and developing new ones. However, when the broader, corporate-enterprise view of the
endeavor was taken, it was clear that the Knowitall representatives should
not be—for the good of the Mineitall Company—encouraging and abetting
the BD personnel in their quest for new conquests.

So, the question is, for whom does Knowitall work? Just because they
were invited to Mineitall by their BD representatives, does that mean that
Knowitall consultants should aid in executing a BD strategy that clearly is
not in the best interest of the greater enterprise? Joe believed that he could
not contribute to a BD effort that was not, in Joe's opinion, in the best interest
of the company.

I can tell you from personal experience that this type of dilemma presents
itself quite often. One of the basics that I would pass along to the reader is to
always be aware of how the entity for which you are providing risk services
fits into the overall structure and strategy of the larger organization. If aid-
ing the smaller entity seems to you to be counterproductive—from the larger
organization's point of view—then you are obligated to at least express in
writing your misgivings. Always endeavor to do the right thing for the overall entity because sometimes those within enclaves of the organization are
blinded by internal silos and are motivated by the reward system.

Be Clear as to "What Is the Question?"

Although I have already offered a brief diatribe on "What's the Question?"
in Chapter 1 (in the section "For Absolute and Relative Purposes") and will
relate how this topic is integrated into the ideal risk process in Chapter 6, it is
worth a mention here because establishing What is the Question? is, in fact,
a fundamental. A journey of a thousand miles begins with but a single step,
and this is that step with regard to RA and risk monetization.

It is not uncommon for each of those who claim to be united to, in fact, harbor divergent ideas concerning the task at hand. For example, a gathering of the major stakeholders in a project might include the project manager, some engineers, some scientists, a commercial analyst, an environmental representative, a financial analyst, a lawyer, and so on. Although these representatives might have physically coalesced and would each proclaim that he or she is present in service of the same project, in fact, each person has a very different project in mind.

A project manager might have the broadest perspective. For that person, the physical, political, schedule, and other aspects of the effort are paramount. This person is most likely to view the project from its inception to its hand off. In contrast, a commercial analyst likely is not so concerned about preconstruction site preparation and the like, but has as his or her concerns market forces, prices, sales, revenue, and similar early project considerations. Construction details, engineering tasks, and so on likely to arise later in the project usually are not part of the commercial analyst's purview. Still different is the construction engineer's perspective. This person is not so concerned about the commercial considerations or the contractual and legal aspects but rather is focused on aspects of the project when construction becomes the primary focus.

Not only does each representative have his or her personal view of just what is the project, but each person has a unique understanding of the temporal aspects of the effort—that is, when it starts and when it ends. For the project manager, the project will begin with the earliest stages and this person's focus on the project will prevail until the project is handed over to another group (perhaps operations) for tending. The commercial analyst is primarily concerned with assuring decision makers that the effort will be commercially viable. These are primarily, but not necessarily exclusively, early-project-stage considerations. Therefore, this person's view of what is the project is very different from that of the project manager. The construction engineer's project starting time is more toward the middle of the project's full life cycle—between site selection or preparation and production start-up phases.

Typically, it is fine for a representative from a particular discipline to harbor a parochial view of the project. However, for the purposes of threat or opportunity identification and for defining mitigation or capture actions, the views of representatives from seemingly disparate disciplines must be as close to being coincident as is practical.

Primarily this is important because decisions and actions taken by one discipline can have salient ramifications for another. Just for example, the commerciality of the endeavor might be predicated upon the use of a particular feedstock grade. This would be decided early in the project life and would be part of commercial contracts. Use of a particular grade of feedstock, however, can have major implications for those who will, much later in project life, design, build, and operate the facility. Unless the specifications are made apparent to all disciplines early in the project life, risk-identification workshops and the like are likely to not take account of such considerations and attendant risks will not be identified.

A simple mechanism to utilize that can promote coalescence of views of the project is to have the project manager—at the beginning of any gathering of discipline representatives such as a risk-identification workshop—describe in detail just what is the timing and scope of the project. This vision of the project should be taken by every representative when considering what might be threats or opportunities. This view of the project's life should not only be related at all meetings, but should be captured in a document that is distributed to all. An agreed-upon vision of the project is essential and is a fundamental.

Be Certain of the Right Level of Support

It seems there are "a million" (OK, maybe not a million) aspects to RA, RM, and monetization that would fit into this "The Fundamentals" chapter. As a practitioner of risk processes, I find it almost unbearably difficult to restrain myself here and to select just a few attributes to emphasize. This is one of those aspects that bubbled to the top.

This risk stuff is hard. It is not that the math or the statistics or the concepts are particularly daunting. Risk stuff is hard for reasons that have to do, mainly, with human nature and with the way modern corporations and other entities are structured.

Risk stuff is hard because it requires people to do things differently. Risk stuff is hard because it embraces subjectivity. Risk stuff is hard because it cuts across silos and requires the participation and cooperation of seemingly disparate disciplines. Risk stuff is hard because everyone has a different view of just what it is. Risk stuff is hard because it often runs counter to the incentive systems. Risk stuff is hard because it often offers a range of "answers" rather than "the number." Risk stuff is hard because those at the top of the house often proclaim that they are in favor of implementing risk processes, but only mean that if it all happens "below" them so that their world does not change. Risk stuff is hard because it forces quantification and documentation of otherwise "assumed" aspects. Risk stuff is hard for a whole host of other equally important and valid reasons.

Did I mention that risk stuff is hard? Most people look at the technical aspects—Monte Carlo analysis, risk identification, and other aspects and decide that it should be relatively easy to embrace and disseminate a risk process in an organization. This perspective often is held by those who occupy the rarified atmosphere in an organization—that is, the organizational leaders.

Nothing sells like success, but nothing leads more quickly and surely to failure than unmet expectations—regardless of how unrealistic were those expectations. Implementation of even a relatively rudimentary risk process in just part of an organization is going to result in seismic shifts in the way people work. It is the responsibility of the risk proponent— you—to identify, evaluate, and communicate these complications to the decision makers. The salient aspects of such a communiqué should not be the technical or process-related methods. Rather, such a presentation should focus on the organizational and cultural ramifications that will result.

It is essential that the "powers that be" are aware of and understand the upheaval that will be commensurate with the introduction of a risk process. Regardless of their bravado on the subject, it has been my experience that they have very little understanding regarding the disruption of the status quo that is required. Support from "the top" is absolutely essential if implementation of a risk process is to be successful. This support has to be informed and real. When low-level managers discover the changes that are required, there will be many an e-mail and call to the upper echelon regarding the hardships that are being foisted upon the lower ranks. If upper management are not stalwart in their belief and backing of the risk process, it will fail to take root in the organization.

So, the bottom line is that it is absolutely fundamental to make sure— before any part of the risk process is attempted—that upper management will stand with you and behind the process when the sure-to-come resistance is generated. Sans support from the appropriate levels in the organization, you are wasting your time.

What I Assume

A review of basic aspects of risk analysis appears in the books referenced at the end of this chapter and I will not repeat in this book those fundamental precepts. However, I thought it would be remiss of me to fail to mention here what I assume the reader understands.

I assume that the reader has a fundamental understanding of the following:

- Monte Carlo analysis
- Frequency and cumulative frequency plots
- Distributions and the concept of "The best distribution you can build is one you can defend"
- Correlation as it relates to the integration of uncertain parameters

Some Definitions

Although I will address the definitions of most critical terms in this section, I utilize the word "some" in the title of this section because I would like to reserve the right to define a few terms and concepts later in this book. Given that the primary focus of this text is the concept of risk monetization and the book is not to be regarded as a primer on the subject of RA/M (see *Risk Assessment and Decision Making in Business and Industry—A Practical Guide*, 2nd Edition, for an expanded explanation of risk fundamentals), I will here relate only the essential aspects of concepts and methods central to a risk-monetization effort.

Risk Concept and Definition

I typically am repulsed and a bit depressed when newcomers to the risk game—and sometimes relatively seasoned folks—begin their risk-related thinking and actions with a spreadsheet, equations, and some sort of distribution builder. RA/M is some about that stuff, but mainly it is not.

Risks are events that can impact the effort—whatever that effort might be. RA/M primarily is about identifying the risks, succinctly defining and describing the risks, quantifying or ranking the risks in some way, identifying threat-mitigation or opportunity-capture actions, integrating the risks, following through on assigned tasks, and constantly and dynamically monitoring the overall risk situation and making appropriate adjustments. As I will relate a bit later and have already addressed in Chapter 1, quantifying and managing uncertainty is at least secondary to actual risk identification, RA, and RM.

As I have put forth in numerous previous publications, I define risk thus:

> *A risk is a pertinent event for which there is a textual description.*

"Pertinent" is an important term in the definition. The term addresses two basic precepts. One is that a risk is something that will impact the effort in a meaningful way. That is, it might be true that in the building of a machine that a mechanic might drop a wrench on the floor of the shop—only to pick it up again and continue his or her work. While it is true that the dropping of the wrench is an event, it is not one that usually is considered to have a significant impact on the cost, timing, or other aspects of the machine assembly. So, the dropping of the wrench would not be considered a risk.

Pertinent also is intended to indicate that a risk is ANY important event. This brings up the sometimes controversial concept of risks being opportunities as well as threats.

I know, I know that traditional training in risk analysis would overtly or by implication indicate that risks are only threats. This, however, is a concept that will serve only those risk analyses that are very limited in scope.

If, for example, we are attempting to estimate the time it might take to accomplish a particular task, we can start with the time typically required to perform the task. Then, the "risk assessment" is, again traditionally, all about what things could happen to cause the task to take longer—longer being worse. A savvy risk assessor, however, will be on the lookout for ways to shorten the time of execution. Minor process changes, rescheduling of major steps, increased efficiencies, and the like are examples of opportunities that could positively impact task-execution time. Therefore, any comprehensive risk analysis should include both the probabilistic events that could increase execution time (the threats) and those that could curtail the time it takes (the opportunities). Only the integration of both the probabilistic threats and opportunities will give a most reasonable estimate of the task time required.

Thus far, we have addressed the pertinent part of *a pertinent event for which there is a textual description*. The next term to be scrutinized is "event."

I could not count the number of times I have heard risk practitioners describe a risk analysis as an exercise in putting ranges around otherwise deterministic (single-valued) values in a spreadsheet. These folks usually employ one of the popular spreadsheet add-on packages that facilitate the building of distributions that will be used in a Monte Carlo analysis. While this practice might (and only might) be part of a RA, it is certainly not the primary practice.

Any RA is an analysis of the pertinent events that might or might not transpire. An event simply is something that could happen that would impact in a significant way the estimated outcome.

Events come in all shapes and sizes. That is, an event can be, for example, a physical thing such as the crashing of a car. An event can be a political thing such as the delay in having granted critical permits. An event can be a temporal thing such as a scheduled task simply taking much less time than was originally estimated (an opportunity). These are all things that could happen—events. Again, these events can be threats or opportunities.

Now, to the last part of the statement *A pertinent event for which there is a textual description*. If you can't tell someone about the event, then it isn't much of an event. Any risk needs to be able to be explained or related to the interested parties. Although I outlined in this in Chapter 2 ("Risk Description—How to Express a Risk"), it is of such gravity that I will briefly address the risk-description subject again.

In Chapter 2 ("Risk Description—How to Express a Risk"), I briefly described how risks should be expressed—in Reason–Event–Impact format. This is an example of—and the preferred method for—how to textually relate a risk. It is commonplace, mainly due to haste, laziness, or simple indifference, to see threats or opportunities expressed with extreme brevity in mind. For example, an opportunity might be documented as "better fiscal terms." Likewise, a threat might be recounted as "poor coordination."

There are myriad primary purposes for completely describing a risk. One is that it might be true that the folks in the room when the risk is recorded understand what is meant by the risk description, but those who have to deal with the risk at a later time are not likely to have the benefit of that context. Another reason to strive for a comprehensive risk description is that it is only through that description that threat-mitigation or opportunity-capture plans can be rationally defined.

Those who are charged with the task of defining or executing the mitigation plan for a threat, for example, need to know whether they are attempting to mitigate the probability of the threat (make it less likely to happen), the impact of the threat (make it less severe in its impact if it happens), or both. This can only be efficiently done if the risk is fully understood.

Thus, the Better fiscal terms opportunity might be expressed in Reason–Event–Impact format as "If the new government regulations regarding environmental-discharge liability are passed by the legislature (Reason), then our company might lower insurance premiums and might be allowed to renegotiate our original contract (Event), thereby lowering our costs and increasing the Net Present Value (NPV) for the project (Impact)."

Similarly, the threat originally expressed as Poor coordination could be revamped as "If our company can't cause the city and county governments to better coordinate their regulatory efforts regarding this project (Reason), then we will certainly be put into the position of trying to adhere to conflicting directions (Event), causing project delay and increased costs (Impact)."

So, I can't stress enough the significance of completely defining the risk—the successful execution of all subsequent risk processes is dependent upon adroit execution of this step. However, that's not all there is to the definition of the term "risk."

Uncertainty and the Confusion that Surrounds It

Associated with each risk—be it a threat or an opportunity—are at least two other things: probability and consequence. These two parameters are, taken alone, easy enough to comprehend.

Probability is the likelihood that the event (risk) will happen. In Chapter 2 (in the section "Probability of Occurrence"), I briefly address the issue of probability and the many views of it. I did a cursory job in Chapter 2 of distinguishing "probability" from "frequency," but I need to amend that discussion here.

Frequency is a view of the past. It is the number of times something has happened. City project planners might be heard saying: "On 6 out of 10 days, there has been an accident at this corner ..." This is the frequency of days there were accidents (but not the number of accidents) in a particular place during the time they were paying attention (10 days). This is a view of the past. This might or might not be a means of predicting the future.

For example, it might have snowed on 6 of 10 days. Therefore, the frequency of "6 of 10" might be a predictor of the number of days we can expect accidents on that corner when it snows, but it is not projecting the likelihood of days with accidents for, say, a year.

Even if we take the "wild card" of snow out of the equation, frequency can still be a poor harbinger of probability. A small sampling can lead to poor predictions of probability from frequency.

Allow me to make example of the too-often-used, but very useful, coin flip. We might flip the coin twice and each time get a "heads." In frequency terms, this is "2 of 2." Is this a good predictor of the next coin flip? Certainly not, because as we all know, the probability of getting a heads on the next flip is but 50%. Only through the recording of the result of many flips will we see the frequency of heads approach 50% of the flips.

Frequencies are typically "hard counts" of things that have happened. Estimates of probability, however, are attempts to be prescient. In the coin flip example cited above, we would strive to assign a 50% probability to getting a heads on the next coin flip in spite of the fact that the previous two flips resulted in heads. This forecast would run counter to the evidence afforded by the frequency information and would have to be based upon what we know about the coin, how it is flipped, and so on. In the case of predicting the number of accident days, we can expect in a year at a particular corner, our prediction of probability would have to take into account the anomalous snow days. If the sample of 10 days of observation were all we were to get, then the frequency data would be most misleading and the generation of a probability of accident days on that corner might have to be predicated on analogy, for example.

One of my favorite anecdotes I like to relate is one which relates to two archrival sport teams. Over the past 60 years, the teams have met in the regional championship match 12 times. Team A prevailed in 10 of the 12 meetings. The last time the two teams met was 8 years ago. Team A won that match. Team A and team B are to meet in a championship match for the 13th time this year. The student bodies from which players are drawn are about equal.

No coaches or team members remain today from the time of their last meeting. The question becomes: Does the historical win–loss record have much if anything to do with a prediction (probability) of which team will prevail this year? Likely not. The quality of the two teams, their current-year win–loss records, injuries, and many other factors have more to do with the probability of winning the match than does the historical record.

Life typically affords you only one shot at things. "Do overs" are rare. I am a proponent of Monte Carlo analysis, but the fundamental insanity of the premise that we will run, say, 1,000 simulations is not lost on me. In real life, we will get just one shot at this, and predictions of things like probability need to be based upon as broad a knowledge base as is practical to assemble. Except in unusual circumstances, counting (no pun intended) on frequency to be a cogent predictor of probability is not good practice.

As I said, the other parameter associated with a risk is consequence (or, impact). The magnitude of consequence is also, most times, a prediction. Sometimes it is not. "If we violate the statute, the fine is $1,000." In the afore-mentioned situation, there is no doubt about the consequence (there might be doubt, however, about the probability)—it is $1,000. However, in many instances, the consequence is a prediction.

Similar arguments apply here as in the prediction of probability. The magnitude of consequences in the past might not be reasonable predictors of future impacts. Also, small samples of consequences (like the two heads in the two coin flips)—just as in the case of probabilities—can be misleading and the estimation of actual consequence might fly in the face of the frequency data.

So, if we are not sure about the probability and/or impact associated with a particular risk, then we must be uncertain (duh!). One of the points to be made here is that we are not uncertain about the risk itself—it is an event that is described in Reason–Event–Impact format. The things we might be uncertain about are, at least, the associated probability and impact.

Not Primarily about Minimizing Uncertainty

In Chapter 1 (in the section "A Focus on Success and Value"), I relate the story of a project manager who revamps aspects of the material trans-port, crane utilization, employee interaction, and other considerations at a building-erection job site. The point of the diatribe is that although the project manager might have been successful in driving down the proba-bility of failure (in this case, an injury accident), that manager has to be relatively uncertain regarding the actual probability reduction that will be achieved—relative to hundreds of examples of job-site experiences doing things the traditional way. So, although the probability of success (no acci-dents), hopefully, has gone up, the uncertainty around the probability has risen significantly.

Another such example might be found in the military arena. For a given weapons system, the military currently experiences a 45% over-the-horizon-first-shot-hit rate (hitting an unseen target with the first shot). They have practiced this process, literally, thousands of times under all sorts of condi-tions and are confident in their hit rate (45%).

A new Colonel has been assigned to oversee the weapons system at the primary training base for the system. He ordered a review of all compo-nents including projectiles, communications equipment, software, computer hardware, weapon hardware, satellite communication, and so on. From the review, it became clear that lag time in satellite-to-ground and ground-to-ground communications and data processing were responsible for the less-than-50% hit rate for the system.

A complete upgrade of the computer systems to dual processors and a replacement of the antiquated software was recommended and undertaken.

The upgraded system has yet to fire its maiden shot, but estimates of the new hit rate are in the neighborhood of 75%.

Before the upgrades were enacted, if an informed soldier was queried regarding the hit rate of the system, he would have answered confidently that it was 45%. Years of experience and tens of thousands of trials back up that estimate. That is, the army was relatively certain regarding the hit rate. Now, things have changed—hopefully for the better. Although the hit rate is now projected to be significantly greater, so is the uncertainty about that hit rate. Having not tried it yet, the uncertainty in the hit rate has substantially increased.

So, what's your point? My point is that the aim of most risk analyses should be to increase the probability of success, to increase the value of the endeavor, or both. Sometimes certainty is "sacrificed" in this process. That is, to be more successful, many times uncertainty has to increase. Therefore, when mitigation actions are undertaken to offset threats or capture actions are enacted to ensure opportunities are taken advantage of, uncertainty typically increases. The mitigation or capture actions are aimed at increasing the success and/or value of the project, but given that these actions are being taken for the first time in the project, the uncertainty regarding the magnitude of the coefficient of success is likely to increase—relative to the initial estimate of success. Aim to increase success and value and let uncertainty fall where it may (it is your "degree of freedom").

There's Uncertainty, and Then There's Uncertainty

This uncertainty thing can get philosophically convolute. I have had many a "discussion" with project people and others regarding just what is meant when we use the term.

I really hate to do it to you, but I'll fall back at least one more time on the coin-flip example. In that example, there is no uncertainty about the probability of getting a heads or a tails—it is 50% (I have actually had people tell me that it might not be 50% because of the chance that the coin might land and stay on its edge—but let's not go there, OK?). Now, 50% is not 100% (profound, huh?). Therefore, if it is not 100%—that is, we are not 100% certain—does it not simply by its nature imply that there is uncertainty? Sure it does. There is a 100% chance that the probability of a heads, for example, is 50% (confused yet?). That is, there is no "range" around 50%.

So, if there is no uncertainty (range around) the probability (50%) why is there still uncertainty in the coin flip? Well, of course, it is because there is uncertainty in the consequence (getting a heads or a tails)—or is there?

Is there any uncertainty about what are the consequences? No, there is not. You will either get a heads or a tails. No uncertainty about that. We are absolutely sure as to what the outcomes will be. Some would argue, then, that there is no uncertainty in the system—and I have had that argument made to me. If that is true, however, then why can't you predict with 100%

accuracy the outcome of the next coin flip? The answer, of course, is because there is what I have come to call (and I don't expect you to follow my lead, necessarily) the "inherent" uncertainty in the system. That uncertainty, to be sure, stems from the fact that there is a range of potential consequences. Of course, it is bimodal, if you will—a heads or a tails—but it is nonetheless a range.

If this bimodal set of consequences didn't exist—like with a two-headed coin—then there would be no range around the probability of a heads and no range around the consequence (a heads, no matter what) and, therefore, the coin flip would not be considered to be a risk. It would then represent an event that is certain to happen and that will have a consequence that is set.

So, this leads us to the unavoidable conclusion that if something is going to be considered a risk, it must have associated with it uncertainty—either in its probability, in its consequence, or in both. In the coin-flip scenario, for example, the argument that there is no uncertainty related to the consequence because we absolutely know the two potential outcomes (no uncertainty about the result being either a heads or a tails) is rejected. By nature of there being more than one possible outcome, there is uncertainty in the system and, therefore, the scenario qualifies as a risk if all other risk criteria are met (like it being pertinent).

Expressing Uncertainty

Although there are myriad ways to capture and relate uncertainty, I am going to focus in this text on the distribution as the primary vehicle of expression. Other methods such as traffic lights, percentages, colored maps, and the like have their place in the scheme of things, but to link with technologies that incorporate expressions of uncertainty—such as Monte Carlo analysis—I had to choose an uncertainty capture or utilization methodology, and I chose distributions as that vehicle.

As the reader likely is aware, there exist many formal names for various distribution types. Normal, lognormal, Beta, chi-square, Weibull, and many other monikers exist for distributions. Each of these, regardless of the type of input required to define it (mean, standard deviation, nu, etc.), is nothing more than an attempt to describe a "distribution shape" which, of course, does not really exist. OK, Glenn, that's an obscure statement—just what are you trying to say?

Any distribution is nothing more than an array of discrete values on an axis. This is more true today with the almost exclusive use of digital computers (as opposed to analog), but we won't go there. In Figure 3.1, I have depicted an array of 15 values on an axis. Typically, in any distribution-building algorithm, many more discrete values will comprise the distribution, but because it is impractical to generate hundreds of points with PowerPoint (the "graphics" package I'm using to produce all of the figures for this book), allow me to be lazy and only use 15 values.

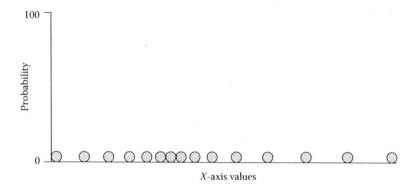

FIGURE 3.1
Scatter of 15 discrete values that comprise a distribution.

An array of discrete values such as those shown in Figure 3.1 is the distribution. Any algorithm that creates distributions (normal, lognormal, Beta, etc.) is nothing more than a means to position the discrete values on the X-axis. Now, I said previously that distributions really don't have shapes, and I stand by that.

"Shapes" for distributions can be created in many ways. A density function can be used to create a "smooth" line (not shown here) that depicts in graphic form the distribution of the discrete values on the X-axis. However, a more popular and pedestrian method is to create a histogram (and histograms are distinct from bar charts) of the point on the X-axis.

If we use the bars of a histogram to express the shape of a distribution, then that shape in part depends upon the number of bars utilized. For example, consider Figure 3.2 in which I have chosen to use just one bar to express the shape of the distribution. If only one bar were used to express the shape of any of the distribution types mentioned (normal, lognormal, Weibull, etc.), all distribution shapes would be the coincident. Of course, the use of one bar is silly, but no sillier than the use of more bars.

Consider the distribution shape depicted in Figure 3.3 in which just two bars are used. It's certainly different than the shape of the one-bar distribution shown in Figure 3.2, but is not much more informative. Now, let's skip ahead in Figure 3.4 to the use of four bars. The shape is different yet, and the shape will continue to evolve until we have one bar for every data point.

So, the point to be made here is that no matter how hair-raising the algorithm utilized for distribution construction, in the end, a distribution is nothing more than an arrangement of discrete values on an X-axis. Histograms and the like are convenient means of attempting to convey to our gray matter just how those discrete values are arrayed on the axis, but any such depiction is only a (usually poor) proxy for the actual distribution of points on that axis.

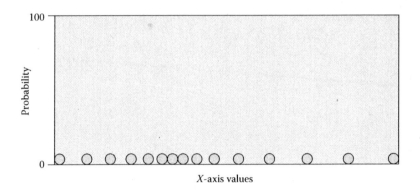

FIGURE 3.2
"Histogram" of the distribution in Figure 3.1 represented by just one bar (class).

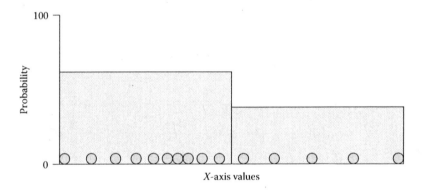

FIGURE 3.3
Histogram of the distribution in Figure 3.1 represented by two bars (classes).

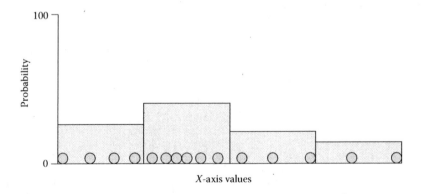

FIGURE 3.4
Histogram of the distribution in Figure 3.1 represented by four bars (classes).

Much of what I am going to relate directly below regarding distribution building has been described in great detail in other books I have penned [see Koller (2005, 2007)], however, I would be remiss in my duties if I did not again relate here some of the more primary concepts.

As espoused in those previous texts, my fundamental premise for distributions is: The best distribution you can build is one you can defend. When I say "defend," I mean that you should have a compelling story, that can be related to anyone who might ask, as to why you chose to use a particular distribution of values.

If your defense of a distribution has anything to do with a mean or a standard deviation or kurtosis or variance or skewness or any other such thing, then you had better beat a hasty retreat and come up with another explanation. Nobody, but nobody wants to hear about a mean or standard deviation or the like when an explanation is being given regarding why a particular distribution of points was chosen.

When a boss or other querying party requests an explanation, what they are really asking is for you to explain the rationale behind your selection of the values you have used and the arraying of those values. For example, let's say you have generated a distribution of commodity price. When the boss asks why you chose this distribution, what he or she does not want to hear is: "Well, because it has a mean of 10 and a standard deviation . . . ." No, what they want to hear is a business reason for the selection of the array of values such as: "Well, we believe that the competition will start up a new production plant in that part of the world next year and so the minimum value of the distribution reflects the maximum depression of the commodity prices we might expect due to the competition. The maximum value for the distribution reflects, in spite of the start-up plant by the competition, the maximum rise in commodity prices expected worldwide due to increased demand for the product in emerging third-world markets and the peak in the distribution reflects where commodity prices are most likely to be next year . . ."

Now, the boss might not agree with your reasons or your selection of values, but at least you have put forward a compelling story that explains your choice of distributions. Again, the best distribution you can build is one you can defend (with a compelling story).

In previous texts, I have explained my distain for the triangular distribution and why it is inherently evil. I will not replow that ground again here, but I will say that the uses of asymptotically tailed distributions are far superior to the use of triangles—especially when it takes no more input data to build such asymptotically tailed distributions as it does to build a triangle.

Let's face it, triangles are not used because they best represent naturally occurring or business-generated arrays of values—they almost never do. They are used because you can enter just a minimum, peak (where the "hump" of the distribution will be), and maximum value and get yourself a distribution. A distribution-building algorithm I have constructed utilizes combinations of minimum, peak, and maximum values along with a "confidence" value to build almost any array (shape) of distribution. However, no matter what algorithm you use to build a distribution, that distribution, when actually used in, say, a Monte Carlo process, will effectively have a minimum and maximum value (even if those values are calculated "on the fly" and not expressly defined) and will have at least one narrow range on the X-axis where there is the greatest concentration of values (the hump or humps in the distribution). Of course, things like uniform, single-valued, and other types of "distributions" can be exceptions.

I always find it amusing that some folks balk at collecting from discipline experts the minimum (P0) and maximum (P100) values and prefer to collect things like P10 and P90 values. Their defense of this position is almost invariably the proposition that we can't know what are the actual "end points." However, they seem to have no problem with coming up with values that are, say, 10% more or less than those points you can't know. In addition, regardless of whether you define the end points explicitly or not, there will be some.

If you are describing with a distribution the distance—in miles—to a star, then selecting a P90 value and having the actual distribution tail "stick out" farther than that P90 value likely is of little consequence. However, if you are working with, say, lawyers and they relate to you the maximum penalty that can be imposed in a sensitive legal case, it is absolutely *not* OK to allow the tail of the distribution to extend beyond that value. Funny, though, how if a person wants to build a triangular distribution, suddenly the capture of minimum and maximum values is alright.

So, I would argue that the best method to utilize is some distribution-building algorithm, which generates asymptotically tailed distributions and one in which you can explicitly define the minimum, maximum, and "peak" (where the hump will be) values. More than one hump (bimodal or more) distributions are common.

I fully realize that I have addressed the issue of "anchoring" in the section "Probability of Occurrence" of Chapter 2. I will here briefly recount some of that same information for two reasons. One is that the subject of anchoring is important and that if anchoring is not avoided, then the discussion above related to distribution shape is moot. The second reason for here briefly revisiting the concept of anchoring is readers use texts like this one mainly as reference books and do not read books like this one from cover to cover as they would a novel. For such readers, I would not want them to forego the important anchoring advice because they skipped over Chapter 2.

When capturing data for a distribution, the fundamental aim is to glean from the data provider a set of objective parameters that capture a best approximation of the real-world uncertainty. Therefore, it is not recommended that you begin by attempting to capture first the peak or "most likely" (where the hump will be) value. It is human nature that once one has convinced oneself of the likely outcome, it is difficult to imagine values that deviate significantly from that most likely value (a phenomenon called anchoring). Therefore, it is recommended that you begin your query with either the minimum or maximum value.

Which value you start with matters. Always start your query with the "good end" of the distribution. For example, if the distribution in question represents revenue, then it is likely that the maximum end of the distribution is the "good" one. Conversely, if the distribution represents costs, then typically the "best" costs are those near the minimum end of the array. People always want to tell you how wonderful it can be.

When capturing data at either end of the distribution, always coincidently record the reasons why that minimum or maximum value is the one to be used. For example, if we are first discussing with a discipline expert the minimum value for a range of labor costs, be sure to record why that expert believes the minimum-cost value is valid. That expert might put forth that they envision a scenario in which they might utilize nonunion labor, that other work for the types of craftsmen being considered will be scant, and that due to the recession, labor prices overall will be at low ebb. Having captured these reasons for the good cost estimate (minimum value), it is your job to use those rationales to "turn the tables" on the discipline expert when discussing the "bad end" or maximum cost value. What if we do have to use union labor? What if the desired craftsmen are in relatively high demand when we start our project? What if the recession eases by the time our project commences and overall labor costs are up? Only in this way are you likely to capture the broadest and most real-world-emulating range of values. Only after you have established the minimum and maximum values should you discuss what is most likely to happen.

Chance of Failure

Chance of failure (COF) is one of those things that seems simple enough, but in reality gets relatively complicated. COF, however, is one of the fundamental tenets of RA, so it warrants a few paragraphs of explanation here. In the following paragraphs, I will relate something about from where we get our estimates of COF.

COF emanates from, at least, two sources. There is failure that is associated with the ranges expressed in distributions and there is failure that is not necessarily associated with distributions. I'll first discuss the latter type. A more complete discussion of COF can be found in Koller (2005, 2007).

Chances of failure are like links in a chain—if any one of them breaks, that's it. For example, consider the "links" shown in Figure 3.5. These links represent the sources of failure associated with a possible car trip.

We would like to take a trip in our car, but there are several things that, if they "go wrong," could prevent the trip from happening. The transmission might be faulty. The engine might not start. The cooling system might not

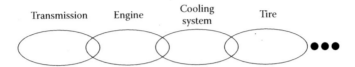

Chances of failure are like links in a chain—if one breaks
(fails), the entire project fails. In this example, the project
is a car trip.

FIGURE 3.5
Chances of failure for a car trip represented as links in a chain.

work. A tire might be flat. The electrical system might have a fault. If any one
of these things is true, then the trip does not happen as planned.

These types of COFs typically are estimated as percentages. That is, there
might be a 5% chance that the transmission might be faulty and a 1% chance
that the engine might not start, and so on. Of course, these percentages might
be expressed as ranges (a 3–5% chance the transmission is faulty). Using discrete values for each COF, the total chance of success (TCOS) is calculated
thus:

$$\text{TCOS} = (1 - \text{COF transmission}) \times (1 - \text{COF engine not starting}) \times \cdots$$

If ranges are used to represent the COFs, then in the Monte Carlo process, on
each iteration, a COF value is selected and substituted in the above equation
to arrive at a range of TCOS values. As will be demonstrated later in this
chapter and book, the TCOS values are instrumental in performing a cogent
risk analysis.

Then, there are COFs that owe their existence to part of the range of a distribution. The easiest way to convey this sometimes erudite concept is with
a relatively simple example. Consider the distributions shown in Figure 3.6.
Each distribution represents a parameter in an oil exploration project. Shown
are distributions for porosity (the percent "holes" in a buried rock that might
contain oil or some other liquid), trap volume (the physical volume of the
buried and porous rock that might contain oil), and recovery factor (the percent of the oil that can economically be extracted from the buried rock).

Buried rocks that might contain oil are called "prospects." Each parameter
related to a prospect, like the three described in the preceding paragraph
and depicted in Figure 3.6, might be represented as a range of values. For
each parameter, the total range of values that might pertain to that variable
is represented by the range of all of the bars in the histogram. For example,
for this type of prospect in other parts of the world, we know from analogy
that the porosity range, if it has any porosity at all (if the porosity is not zero),
can be from 1% to 20%. The range for trap volume might be from 500,000 to
10,000,000 acre feet, and the recovery factor can range from 3% to 60%.

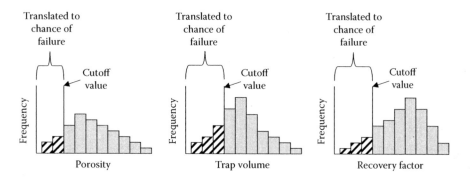

FIGURE 3.6
Depiction of how chance of failure can be derived from the combination of a distribution and a "cutoff" value.

However, we also know that even if we have acceptable trap volume and recovery values, if the porosity is less than 5%, then the flow rates of oil will be too low and the prospect will be a commercial failure. Similarly, even if the porosity and trap volume are acceptable but the recovery rate is unacceptable, then the prospect will fail because not enough hydrocarbons will be brought to the earth's surface and, therefore, to market. In our example, for porosity, the 5% coefficient is termed the "cutoff" value. More about this will be discussed in the section "Collect a Cutoff Value" of Chapter 6.

Similar "failure" ranges exist for trap volume and recovery factor. For each parameter in Figure 3.6, the failure range is shown as the crosshatched bars. In keeping with the "chain" concept shown in Figure 3.6, if the prospect in question harbors a value in the "failure range" (hatched bars) for any one of these parameters, then the prospect will abjectly fail.

One of the primary objectives of probabilistic prospect analysis is to generate a range of potential hydrocarbon (gas or oil, for example) recoverable volumes. Any energy company is in the business of making money. Therefore, the volumes of hydrocarbons extracted from the earth need to be done so in an economically feasible manner.

It is typical that one or more hydrocarbon volumes is passed through a series of subsequent modules on the way to a financial model. Modules between the volume-generating process and the financial engine can be drilling, facilities, transport, costing, and others. In the end, the financial module will generate a prospect-related full-life-cycle estimate of the value of the prospect—typically an NPV, Internal Rate of Return (IRR), or another such financial metric.

It stands to reason that any hydrocarbon volume that is passed to subsequent modules for, eventually, financial analysis should be a volume that would be successful with regard to "below-ground" considerations. That is, that volume should represent a successful porosity, trap volume, recovery factor, and so on. Other "above-ground" considerations (risks) such as commodity price, facilities costs, transport costs, drilling costs, and so on will be added to the mix, but it is essential that the volume that is being considered is not one that would have failed for below-ground reasons. More simply put, why would you want to financially evaluate a prospect that was a failure to start with? You wouldn't.

So, let's assume that we have available a Monte Carlo model that can integrate all of the below-ground parameters (porosity, trap volume, water saturation, recovery factor, and more) to produce a range of hydrocarbon volumes. Let's further assume that we put into this Monte Carlo model the full distributions shown in Figure 3.6—that is, both the white and hatched bars for each parameter.

A cumulative frequency plot such as that shown in the top plot in Figure 3.7 will result. This cumulative frequency curve might be the result of 1,000 Monte Carlo iterations—that is, the curve is comprised of 1,000 calculated hydrocarbon volumes.

Note that the cumulative frequency curve begins at its left end at 100% on the Y Probability axis. Now, remember that according to the "links in a chain" concept for COF that when calculating any of the 1,000 points that comprise the curve, if even one value was randomly selected from the hatched-bar area for any one of the input parameters, that volume will represent a prospect that—in real life—would fail. I challenge the reader to pick

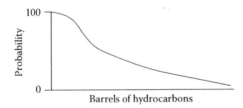

Cumulative frequency curve resulting from Monte Carlo integration of the entire range of the distributions shown in Figure 3.6.

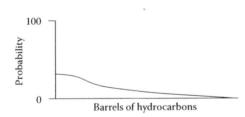

Cumulative frequency curve resulting from converting the striped bars in Figure 3.6 to chance of failure (the Y-axis intercept in this plot) and using the bars to the "right" of the cutoff in the Monte Carlo process.

FIGURE 3.7
Top cumulative frequency curve resulting from Monte Carlo sampling from the entire range of values such as those (and more) shown in Figure 3.6. Bottom cumulative frequency curve resulting from sampling from the solid-shaded bars in the distributions shown in Figure 3.6 (and more) and using the number of X-axis values represented by the striped bars to represent chance of failure. Chances of failure are integrated to determine the Y-axis intercept of the cumulative frequency curve.

out on that curve the failures from the successes. The failures do *not* exclusively represent those values at the "low" end of the curve.

It is perfectly reasonable to expect that on one Monte Carlo iteration, we would have selected a giant trap volume, a great recovery factor, but 4% porosity. The less-than-5% porosity would cause the prospect to fail in real life, but the volume represented by that failed prospect could be near the "right" or "large volume" end of the curve.

So, if we used a curve like this one from which to select our hydrocarbon volumes to be financially assessed, how would we know which values represented successes and which represented failures? Turns out, without analyzing the specific input data for each of the 1,000 points and also assessing how those values interacted with one another in the Monte Carlo model, it is impossible to tell—just from the magnitude of the volume—whether any given volume on that curve represents a success or failure. Well, you say, what can be done about that? The COF concept, that's what.

Each of the histograms shown in Figure 3.6 has hatched bars that represent the failure range of the parameter. A well-designed distribution-building algorithm can easily calculate the percent of the area of the distribution that is represented by the failure values. A less sophisticated algorithm can simply count the number of X-axis values in the failure range and express the failure as the percent of the total values on the X-axis. The least sophisticated method is to estimate—by eye—the percent of the histogram that is represented by "failure bars."

When a percent failure is calculated or estimated for each input parameter, the TCOS (total chance of success) is calculated as indicated in a previous section:

$$TCOS = (1 - COF \text{ porosity}) \times (1 - COF \text{ water saturation}) \times \cdots$$

In our Monte Carlo model, *only* the white bar range—that representing the "success" range for each parameter—is input to the Monte Carlo model. The TCOS calculated is used as the Y-axis intercept for the cumulative frequency curve. This process results in a cumulative frequency curve like the one shown as the bottom plot in Figure 3.6.

Note that the shape and range for this curve is different from the one above. Also, every point on this curve represents a successful prospect—one that could with confidence be put through the financial analysis process. If 1,000 Monte Carlo iterations were performed, the curve is comprised of 1,000 successful hydrocarbon volumes. In addition, the probability of attaining any one of the volumes is strikingly different from that indicated in the top curve in Figure 3.6. The bottom-curve probabilities associated with any given volume match the probabilities associated with those volumes experienced in the real world. This use of COF is just one of a series of such applications that will be highlighted by examples in later chapters of this book.

Mitigation and Capture

Although I took a cursory swing at discussing threat-mitigation and opportunity-capture concepts in Chapter 2 (in the section "Threat-Mitigation or Opportunity-Capture Action"), I did not in that section address the actual definition of the terms "mitigation" or "capture." I will here do so.

As discussed in Chapter 2, one must always be aware of just what is being mitigated or captured. With regard to mitigation, just for example, one should be cognizant of whether mitigation steps are being taken to mitigate the probability, the impact, or both. In Chapter 2, I use the example of car-building versus airplane-building companies and their very different views on just what can effectively be mitigated. The same lessons apply to opportunity capture.

A glaring example of risk-related miscommunication having a significant impact on the fortunes of an organization relates to the precise meaning of the term mitigation. In some circles, mitigation means: What you do to fix a problem after the problem has materialized. I can't tell you the number of safety manuals I have seen where the "mitigation action" for a spill, for example, is defined as how to safely clean up the spill.

An experienced risk assessor and manager will always define mitigation as an action that we can take "now" in an attempt to assure that the threat will never materialize. Similarly, the term capture relates to actions to be taken now that will better assure that an opportunity will be captured.

I always relate this to the proverbial long-distance trip in the car. You might note that prior to your trip, some of the lug nuts are loose on one of the car's wheels. Mitigation of this threat does not mean fixing the body damage and replacement of the tire and wheel when, 100 miles down the road, the wheel comes off the car. In proper risk parlance, mitigation means having identified the threat, doing something about it now so that the threat will not materialize. The mitigation action might mean going down to the local tire store and for a small fee having them tighten the lug nuts before embarking on the drive.

Differentiating between the two views of mitigation might seem trivial. However, I can assure you that it is anything but trivial when a risk register containing dozens of threats and opportunities has an equal number of identified and defined mitigation actions—half of which are "fixes" to the problem after it might happen (what to do about the spill if we have one) and half of the actions are properly defined as actions we should take to cause the opportunity to be captured or the threat to be avoided. Attempts should always be made to avoid threats and to capture opportunities and risks for which fixes are defined (for after the risk materializes) are those that might have been avoided altogether, but were allowed to happen because of miscommunication regarding the term mitigation.

Because I have devoted a large portion of a book to the subject of the incentive system and how it impacts the take-up of risk processes in any organization [see Koller (2007)], I won't here laboriously belabor on the subject. However, because an organization's incentive system exerts such a powerful influence on the actions of people, I feel I would be remiss in my obligation to the reader to here eschew the subject and its relation to threat mitigation and opportunity capture.

As you will see later in this book, an early-in-the-RA-process step is to conduct a risk-identification workshop. At this workshop, threats and opportunities are identified and captured in some sort of risk register (see Chapter 2 for much more on risk registers). Associated with each risk, a threat-mitigation or opportunity-capture action and cost are defined.

It is true that most projects go through phases. For example, a construction project might go through the phases: Concept definition, negotiations, site selection, site preparation, building design, construction, and operation. If you were to hold a risk-identification workshop early in the project life cycle, you might be doing that during the concept-definition phase. The people responsible for this phase of the project will largely not be the same individuals who will take responsibility for subsequent phases. However, it is on the concept-definition people's watch that the foreseeable threats and opportunities—as best as can be done at that early time—for the entire project were identified and recorded in the risk register. Mitigation and capture actions and costs also were recorded.

Again, mitigation is defined as a step that can be taken now in an attempt to ensure the threat never materializes. Similarly, a capture action is one to be taken as soon as possible to ensure an identified opportunity is captured. The operative word in that definition of mitigate is now.

People responsible for the initial phase of the project are rewarded for getting their part of the project done on time and on budget. Now they have identified a whole host of actions (threat mitigation and opportunity capture) that should be enacted just as soon as is practical—possibly on "their watch." These actions will take time and money. Hmmmm .... As the purveyor of the risk process, what words could come out of your mouth that could cause the Phase-1 people to want to spend much time and copious amounts of cash to, for example, mitigate threats that might or might not ever happen, and, certainly will not happen on their watch?

"Be a team player" and "Take one for the team" are hardly promotions that will cause these folks to exceed their allotted time and budget so that others in the future will have an easier go of it. This is especially true when these folks are being rewarded for coming in on time and under budget, for example. And besides, these risks might not actually happen at all, so where's the incentive to do anything about them?

So, just because threats and opportunities have been identified and recorded in the risk register and associated mitigation or capture actions have been defined, fundamental organizational aspects will need to be addressed if it is to be expected that threat-mitigation or opportunity-capture actions are to be taken seriously. The incentive system has to be changed.

Monetization

It is true that many of the common decision-making metrics in business today relate to money—such as NPV, IRR, DROI (discounted return on investment), ROCE (return on capital employed), and so on. It also is true that the dictionary definition of the term "monetize" relates directly to money. However, in the context of RA and RM, the term monetize does not necessarily have to link directly to money.

In the pages of this book, the term monetize will be much more liberally interpreted. When I use the term "risk monetization," I will be referring to the translation of a threat or opportunity into almost any type of impact on the perception of project value or on a critical decision. The translation of a risk can be purely conceptual, qualitative, quantitative, or any combination thereof.

Getting the risk to be reflected in the perceived "worth" of the project or to cause the risk to be considered in decision making is the point. An example of a purely conceptual translation of a risk into an impact on decision making might be the decision of which route to take to a desired destination.

It might be that there are two routes—A and B—to a destination. Route A is shorter and more direct than route B. However, as you reach the intersection at which you will have to make the A or B decision, someone in the car tells you that route A has some low spots in it and recent rains might cause those spots to be impossible. The low spots and rain are threats—or risks—to the decision to take route A. The intersection is fast approaching and there is no time to resort to probabilistic calculations. The information is considered and a decision is made.

In this book, such a conceptual process is considered "monetization" simply because a risk was identified, considered, and had a material impact on the decision. A higher level of monetization complexity is that of qualitatively considering and translating the risk.

Qualitative risk monetization is somewhat similar to conceptual monetization with the added caveat that categories typically are involved. Going back to our route A or route B example, you might be planning a trip. Sitting at the kitchen table, you learn from an internet search that route A has a *high* likelihood of having significant construction at the time of your future trip as opposed to route B's *low* expectation of construction delays. However, there is a *medium* likelihood that a town along route B will be celebrating a centennial and traffic backups are likely. There is a *low* expectation of town-related traffic congestion along route A.

In this example, high, medium, and low categories for threats and opportunities are created and utilized in the decision-making process. This type of monetization is common in, for example, sports events or competitions that are judged.

Quantitative monetization is just what it sounds like and will be the primary monetization exemplified in this book. Because many quantitative scenarios will be related in subsequent chapters, I will say here that this type of monetization involves the translation of risks into numbers. These quantitative data are typically utilized in some sort of equation or set of equations to calculate a result or a range of results. Such results rarely give the decision maker the answer, but typically serve as one of the decision-making criteria. Much more on this subject will be related in Chapter 6.

Suggested Readings

In addition to books and articles cited in the preceding text, this list includes other material that the reader might find helpful and relevant to the subject matter discussed in this section of the book.

Finkler, S. A. *Finance and Accounting for Nonfinancial Managers.* Englewood Cliffs, NJ: Prentice Hall, 1992.

Gentle, J. E. *Random Number Generation and Monte Carlo Methods.* 2nd Ed. New York: Springer-Verlag, 2003.

Jaeckel, P. *Monte Carlo Methods in Finance.* New York: John Wiley & Sons, 2002.

Koller, G. R. *Risk Assessment and Decision Making in Business and Industry: A Practical Guide.* 2nd Ed. Boca Raton, FL: Chapman & Hall/CRC Press, 2005.

Koller, G. R. *Modern Corporate Risk Management: A Blueprint for Positive Change and Effectiveness.* Fort Lauderdale, FL: J. Ross Publishing, 2007.

Meyer, H. A., ed. *Symposium on Monte Carlo Methods.* New York: John Wiley & Sons, 1954.

Nijhuis, H. J., and A. B. Baak. "A Calibrated Prospect Appraisal System." *Proceedings of Indonesian Petroleum Association* (IPA 90–222), 19th Annual Convention, October 1990: 69–83.

4

Spectrum of Application for Monetization

Human nature is a funny thing—I'll get to that in just a bit. In my career I have taught a lot of classes. I tend to teach classes in which general principles are conveyed. Therefore, unless the class is aimed at a specific discipline, it is not unusual for such a class to be comprised of representatives from a broad spectrum of areas of expertise.

When designing such classes, I work into the proposed agenda myriad hands-on exercises and examples. Of course, each example is predicated on the principles and precepts of a particular discipline. For instance, I might use a large list of environmental clean-up projects to make an example of how risk processes can be utilized to perform ranking of projects in a portfolio of projects.

Human nature is a funny thing for many reasons, but just one of those reasons is that it never fails that when I use a particular discipline to make a point regarding risk analysis—like the environmental projects to exemplify risk-based portfolio analysis—someone who has expertise in a different discipline pipes up and says something akin to: "Yeah, that's a great example for environmental analysis, but that approach would never work in an effort to rank construction projects ..."

Although when teaching a class I unrelentingly strive to convey risk principles that can be nearly universally employed—that is, they can be utilized in projects emanating from nearly any discipline—it rarely fails that at least a portion of the population of the class cannot comprehend the translation of the risk application in a class example to their own area of expertise. Ergo, I take the time and space here to document the universal nature of the art and science of risk analysis and risk monetization.

There is hardly space in any book to address specifically all project types. I will here make example of just a few of the more ubiquitous varieties of projects.

Acquisitions

As you might guess, risk processes—especially risk monetization—play an important role in the decision by a company to acquire another business entity. Of course, the usual financial and commercial considerations apply.

However, it seems that the parameters that can be captured in a spreadsheet as line items are only the opening salvo regarding the list of considerations that should be risk assessed and monetized.

Even relatively huge acquisitions that look good "on paper"—that is, according to the spreadsheet analysis—can be ultimately torpedoed by risks that tend to be overtly overlooked in the typical spreadsheet process. For example, it is rare to find a spreadsheet line item called "Clash of Cultures." I will draw on my own experience to relate to you an example of this.

Clash of Cultures

Some years ago, I was associated with a large company—which shall remain nameless—that decided that they would like to acquire an energy company. With regard to vertical integration, this proposed acquisition seemed logical given that the petroleum products discovered, produced, and refined by the energy company would serve as feedstock for the acquiring company.

The acquiring company sold "finished products" to other companies and to the public. Development of a new product by this company was a decision not lightly or easily taken. Before a new project or product could be launched, a very high degree of assurance was required regarding the usual parameters such as initial market penetration and share, cost projections, revenue generation, intellectual property, market growth, inability for others to develop copycat products, and the like. Probabilities of success in the 80% range or better were desired for each of the aforementioned issues.

Contrast this need for assurance with the "high-risk" culture of any energy company involved in exploration. First, exploration for hydrocarbons (oil and gas) is expensive. It is not uncommon for a well-drilling program for an individual prospect to cost millions of dollars. Furthermore, it is typical that a true wildcat prospect (a proposed drilling site in a substantially unexplored part of the globe) sports a chance of commercial success of less than 20%. The reality of this exceedingly low probability of success for any individual project is that many millions of dollars will be spent on projects that will fail much more often than not. The relatively rare successful project has to produce sufficient revenue to cover the cost of the project and the costs associated with many of the failed endeavors.

Such is the world of exploration in an energy company. These types of risk profiles are exceedingly common in the energy-exploration world and in other venues such as drug development by pharmaceutical firms. In such drug-generating companies, huge capital expenditures for individual drug development are commonplace. Of a portfolio of, possibly, dozens of drugs, only a few are successful (if the drug companies are lucky!) in navigating the complex maze of development, trials, FDA approval, and other hurdles to become a drug that ultimately reaches the intended recipients. Just like the prospect-development scenario in an energy company, the few successes have to pay for the many, very expensive, failures.

Having an earth science background, even I could see—when working for the need-for-high-assurance acquiring company—that there was an impending clash of cultures. The cross-every-t-and-dot-every-i culture was about to attempt to merge with the "cowboy" attitude toward business. Although likely not a line item in the acquisition-assessment spreadsheet, this type of "soft risk" can easily be monetized.

Several equally valid approaches could have been taken. First, a risk-identification workshop might have been held with representatives of management from both companies. Such a workshop should have been facilitated by a disinterested party—that is, a consultant experienced in conducting such workshops.

Any seasoned workshop facilitator would methodically explore cultural issues. Tolerance for risk (risk utility as it is termed in the business) would have been on the list of items to be discussed. I would be amazed if the tolerance-for-risk issue would not have bubbled to near the top of threats to successful merger.

Adhering to the Reason–Event–Impact format for expressing risks, the experienced facilitator would have insisted that the risk be recorded something like:

> If high assurance for sanctioning projects is required by the acquiring company (Reason), then entire portfolios of high risk projects in the acquired company might not be funded (Event) which could lead to the choking off of feedstock for the acquiring company which will in turn significantly retard production and have a major negative impact on revenue (Impact).

Once identified and expressed as above, such a threat can be monetized by multiple means. Just one route to monetization would be the relatively mundane process of assigning a probability to the threat and a range of monetary consequences. The goal might not initially be to integrate this threat into a financial model, but rather to prioritize the risk—relative to other threats and opportunities—in the risk register.

To accomplish this, the risk-identification-workshop consultant might glean from select participants, through facilitated conversation, the range of probability of materialization of this threat. Given that it is imperative in the acquiring company that a high degree of probable success be assigned to any high-cost project to be sanctioned and given that most of the high-cost projects in the acquired company's exploration portfolio are multimillion-dollar projects with low chances of success, the likelihood of this problem rearing its ugly head is high. Let's say the participants settle on a range of 90–100% likelihood.

Next a dollar value can be assigned. In this instance, such a value should not be assigned to the cost of one or more of the acquired-company's projects, but to the loss of acquiring-company revenue if feedstock for its products is curtailed. Such losses could easily range in the millions to tens of millions of dollars over a specified period. Simple multiplication of a selected (from the range of 90–100% likelihood) probability by a selected monetary amount (in the millions of dollars) would yield a financial metric that could be used to rank this threat against other threats in the risk register.

Hopefully, the list of threats would not contain too many other risks with the potential for negative impact as exhibited by this threat. Therefore, this threat would "bubble to near the top" of the list of threats and opportunities. Such prominence in the list would then be followed by inclusion of the probabilistic impact of this risk in the financial spreadsheet calculations. How this might be accomplished will be the subject of subsequent chapters of this book. As it turned out in real life, the merger of the two companies did not stand the test of time and cultural clash was one of the major deciding factors.

Weakest Link

In my definition of monetization, I mention that monetization need not strictly mean that a threat or opportunity is translated directly into money. So, I will here relate another personal real-life experience that will combine acquisition with a nonmonetary monetization of a threat.

I was once involved in a gargantuan project that would require the acquisition of right-of-way for thousands of miles. As you might imagine, such an endeavor would require gaining the right-of-way from numerous counties, states, provinces, and lesser political entities.

Larger political factions typically are motivated by potential revenue. Tariffs, royalties, rentals, and other such potential remuneration weigh heavily in the decision regarding granting of right-of-way. With smaller political entities, however, it is often difficult to imagine the primary motivating factors.

Privately held lands, for example, might need to be accessed. Cash generated from mineral rights on such lands, for example, spread across a relatively small population might diminish the attraction of possible royalties or tariffs associated with the granting of right-of-way. In addition, historical less-than-stellar treatment of the small groups of individuals by businesses and governmental bodies might color the decision to cooperate. In short, any number of atypical negative inducements might impact a right-of-way decision.

Prior to sinking significant capital into such a project, any business worth its salt is going to want to ascertain what is the chance that the project will be successful. This is the Total Chance of Success (TCOS) I discussed in Chapter 3. TCOS is calculated by combining the chances of failure (COF) associated with each parameter to which a COF might apply.

As related in Chapter 3, COF are like links in a chain. The chain is only as strong as its weakest link. Failure of any link will cause abject failure of the entire enterprise. For example, as detailed in Chapter 3, a potential car trip can fail to materialize due to independent failure of a tire, the cooling system, the engine, the transmission, and the like. It matters not how great all other elements are if one of the parameters fails (flat tire, inoperative cooling system, etc.).

And so it was with this project. When required to combine all of the COF for all of the disciplines (engineering, environmental, financial, commercial, health and safety, and the like) to calculate the TCOS for the project, acquisition of privately-held-land right-of-way certainly was one of the considerations.

Before the calculation of TCOS, initial talks and some negotiations had been carried out with various private owners. Not obtaining a right-of-way from any single owner might not have necessarily caused failure of the project because, at great expense, the pipeline could be routed around any given barrier. However, failure to obtain the right of way of multiple and geographically contiguous private owners could have scuttled the project.

So, monetization of this political threat might first be expressed as a series of COFs that, when combined with all other COFs from other disciplines (see Chapter 3 for instructions on how to do this), would result in the project's TCOS. This is monetization of the political threat without translating the risk directly into money. However, translation into cash might also be relevant.

Any financial model should probabilistically determine how many private-owner failures the project might encounter. If, on any given iteration of the Monte Carlo model, the calculations encountered just one failure, then the financial impact of routing the right-of-way around the private lands might be imposed. In a subsequent iteration, however, multiple failures related to contiguous lands might cause cessation of the financial calculation because the model had encountered a scenario in which the entire project failed abjectly. In this situation, monetization of the risks is translated into negative monetary impact when the model encounters one or more noncontiguous parcels of land across which we fail to get right-of-way.

Divestitures

My tale relating to divestitures is not so much about monetization, per se, but addresses the point of the failure of companies to recognize, and, therefore, account for (monetize) the fact that it is almost impossible to push risk and accompanying responsibility off onto another business entity. More on this topic later, but let's first briefly look at divestitures in general. As is true for so many topics upon which I touch in this book, there is not nearly sufficient room in these pages to adequately address all of the nuances of divestiture.

Divestiture today is a significant element of corporate restructuring. It typically involves the sale of a company segment for the purpose of lowering costs or aligning the organization with a new business strategy. Although it might seem intuitively obvious that a company should in some manner rid itself of unprofitable segments, the divestiture strategy should only be undertaken if the *value* of retaining an underperforming part of the corporation is less than the *value* of divesting the entity. More on value later.

Divestiture has myriad guises. Common are equity carve outs, spin offs, tracking stocks, and other flavors of divestiture. It is beyond the scope of this book to delve into the differences in divestiture types, but suffice it to say that just some of the distinctions involve how liabilities are handled, how assets are allocated, how debt is distributed and how shareholder equity and voting rights are managed. Today, in tough economic times, the tax-free spin off is popular for many reasons including that it can lower the debt of the divesting company. Regardless of the type, divestiture is proving to be a favorable strategy for improving the competitive nature of a company, enhancing profitability, and for focusing on core competencies.

A Focus on Value

In the above paragraphs, I have alluded to the concept of "value" several times. A focus on value is a primary tenet of the risk-monetization process. I am not here suggesting that some measure of value is not considered in some divestiture decisions, but clearly it is not a major element in many documented instances.

I have been witness to divestiture decisions that were based primarily on the low profitability of a segment of a corporation. For example, part of a company that is earning less than the cost of capital might become the target of a divestiture. When corporations are answering to stockholders, pressure to rid the business of the seemingly underperforming segment can be intense. If a perceived bump in stock price would result from the divestiture, the parent company's management will be under some pressure to act.

This is where risk monetization and the value concept enter the picture. Let me make an example of an underperforming Logistics Department (LD) of a company. For the reasons listed in the previous paragraph, the parent company might be under pressure to spin off the shipping part of the business—the logistics arm that gets products from point A to point B. While it might initially seem reasonable to divest itself of the LD, a risk-based analysis of value might dissuade decision makers.

In a risk-identification exercise on the topic of divestiture of the LD, threats and opportunities such as those below might be delineated.

> If the currently underperforming LD is divested (Reason), the now guaranteed availability of shipping could be subject to forecast high demands for cargo ships (Event) thus causing delays in product deliveries and a reduction in project Net Present Values (NPV) (Impact).
>
> If the currently underperforming LD is retained (Reason), our current oversupply of shipping volume could be contracted to other companies to meet the forecast high demand for shipping (Event) thereby increasing the profitability of the corporation in the long run (Impact).

A threat is first described above. In the simplest monetization exercise, a range for the probability of the forecast high demand for shipping would be established (percent probability), a distribution generated for the magnitude of the high shipping demand that would cause our company to be without enough contract ships (in whatever units that make sense—tons of shipping, for example), and a range created for the magnitude of $ lost per day due to the unavailability of shipping. If the value metric of the analysis is NPV, IRR (Internal Rate of Return), or some such time-series-cash-flow-based metric, then such probabilities, magnitudes, and costs might have to be allocated over specific time periods in the cash-flow time series.

An opportunity is described above as the second risk. In this case, the simplest risk-monetization process would require a ranges to be established for potential oversupply of shipping (in, for example, units of tons), for the percent probability that the oversupply will materialize (in units of percent probability), and for the range of revenue (dollars per time period) to be realized in the near future. Again, if the value metric is NPV, IRR, or some other measure based on discounted cash flows over time, the values in the afore-mentioned established ranges might have to be distributed over specific time periods.

In the end, the integration of the two risks described above into the calcu-lation of near-term corporate profitability might dissuade decision makers from spinning off the currently underperforming LD. It all starts with not having a knee-jerk reaction to short-term profitability projections, with hold-ing a risk-identification workshop to identify and describe the threats and opportunities associated with divestiture, monetizing the risks, and incor-porating those monetized threats and opportunities into the long-term value projection for the parent company.

Pushing Away Risk

One reason that corporations might consider divestiture of a current-company segment is that the parent company would like to indemnify itself from threats posed by the ownership of the seemingly offending arm. That is, they would like to push away the threat of, say, potential future litigation onto a divested company.

First to consider is the reality of "spinning off" a company. It is not unusual in such transactions to have the parent company remain contingently lia-ble for existing contracts that are inherited by the spin off as well as future related contracts. The parent company often is still liable—to some degree—for the performance of the newly created business. It also might be true that the parent company will be liable, such as acting as a guarantor, for credit extended to the spin off. So, the illusion of a "clean split" with a part of the business and with associated risk is many times just that—and illusion.

For example, a parent company might be comprised partly of a mining subsidiary. Parent-company management have recently witnessed many legal actions against mining corporations for heavy-metal contamination in soils left behind from past mining operations and upon which new housing developments have been established. Clearly, the parent company would not like to be embroiled in litigious goings on, so they might consider divesting themselves of the mining operations by spinning them off as a separate company.

Clearly, the usual business calculations should be made including the cost and profitability of the mining arm, future revenue potential tied to product demand, and so on. A company might actually go as far as to monetize risks such as:

> If tailings from our mining operations are used in residential or commercial land development projects (Reason), heavy-metal contamination of property and persons could result (Event) causing legal actions to ensue which can result in significant fines and other liabilities (Impact).

Such a risk might be monetized (probabilities, magnitude of contamination, and magnitude of liability) and incorporated into the divestiture value estimation. That' is all to the good. However, it has been my experience that it is absolutely not the end of the road.

I have realized that regardless of the indemnifying nature of the contracts that define the divestiture of such entities as the mining arm of the company, such assurances against future litigation are not worth much. I have actually witnessed a case in which tailings from a pile near a mine were deemed to be contaminating—by water runoff—the surrounding terrain. More than half a dozen different companies had owned the mining operations since the early 1900s. Each sale of the operations included indemnifying verbiage intended to protect the seller.

When it came to light that the tailings pile—and it was no small pile—was offending, the prosecuting legal staff took a straight forward approach. They measured the height of the pile. They identified all past owners of the mining operations. They determined how long each company had owned the operations. They apportioned a volume of the pile to each previous owner according to how long they had owned the operations (i.e., if one company had owned the operations for, say, 50% of the mining time, then they would be apportioned 50% of the tailings pile and, hence, 50% of the impending liability).

Cleanup costs of the surrounding now-contaminated terrain were staggering as were the costs of getting rid of the pile. Regardless of the indemnifying contracts that reached back more than three quarters of a century, the prosecuting legal firms successfully, in most cases where the offending entity still existed, extracted at least some funds from the companies.

OK, Glenn—nice story, but what's your point? The point is that when considering a divestiture of an arm of the company, the divesting company should include in its risk-impacted value calculation the potential future costs associated with the divested arm. If the parent company is divesting the entity with an eye toward averting future litigation, it has been my experience that there is no such thing as completely laying off risk onto a spin-off company. Monetization of the risks should be done just as though there was little or no indemnifying impact associated with the divestiture. It is just a fact of the world we have built for ourselves.

Environmental Projects

Monetization of risks is especially applicable to projects that involve environmental efforts. Such projects might include the estimation of potential contamination from a proposed project, environmental assessment of baseline contamination (i.e., what "contamination" has already been provided by nature), evaluation of contamination resulting from man-made operations, modeling of contaminant transport, decommissioning and cleanup of past operations, and efforts of similar ilk. In this section, I'll make an example of the last of the aforementioned project types.

I once served as the "risk guy" on an environmental project that involved the estimation of decommissioning and cleanup costs for a huge geographic area. Left-behind equipment and facilities, liquid and solid contamination, hundreds of miles of raised gravel roads, and other issues prevailed (those of you who are familiar with this part of the world will see right through my thinly veiled attempt at obfuscation).

Fundamental was the question: How much money should a company set aside for future remediation efforts in this exceedingly large area? This might seem a simple query, but right off the bat, so to speak, it got very convolute.

First, the tens of potential clean up sites were scattered over thousands of square miles. In addition, the types of sites were of a wide spectrum including staging sites, roads, operation sites, transport facilities, production facilities, storage sites, housing and other associated living quarters, and so on. To complicate matters, various sites had been owned, over many years, by numerous companies—some of whom were more environmentally responsible than others.

An ultimate confounding factor, however, was the fact that multiple political entities—local villages, state regulatory agencies, state government, federal agencies, and others all had various views of just what it meant to "remediate" any given site. For example, with regard to raised gravel roads (roads wide enough for two large trucks to pass one another in opposite directions and raised several feet above the surrounding terrain), local villagers might view such roads and large raised gravel pads as essential means of transport and "high and dry" land that could be put to use. Other agencies—especially influential environmental activist organizations—were of the opinion that if they could fly over the area and see any trace that you had once had operations there, then you were not finished with your remediation effort. Other political entities held views of remediation that lay somewhere between these extremes.

At any given site, it was difficult to impossible to know which political stakeholder would prevail and have "sign off" power regarding whether or not the company had met remediation standards set for that site. As you can imagine from the information in the previous paragraph, the list of remediation steps emanating from each political entity were very different and, therefore, represented an exceedingly wide range of steps and costs associated with each of tens of localities.

Needless to say, a monstrous effort ensued. Contractors were hired to visit each site and to collect the appropriate data. In addition, representatives from the company had to meet with each stakeholder and glean from them—on a site-by-site basis—just what remediation steps would be required. An equally monstrous Monte Carlo model was constructed by lucky me.

As put forward in Chapter 3, risk monetization is the translation of a threat or opportunity into almost any type of impact on the perception of project value or on a critical decision. In this case, the "threats" were the remediation steps and associated costs. Therefore, once the list of remediation steps was determined for each site for each stakeholder, a range of costs was assigned to each step. For example, at Site A, Stakeholder A might require that all local pits be drained and filled. At Site A, the associated cost might range from $100,000–$500,000. Stakeholder B, however, might require only that water in pits be tested to determine whether any specific contaminants—if present— exceeded specific concentration limits. If no contaminants were found to be in concentrations greater than the specifications, then the pits could be left as is. A cost range of, say, $20,000–$200,000 might be established. And so it would go for each site and for each stakeholder at that site.

Ranges of probabilities for most steps also needed to be ascertained and represented as distributions. This was true for most remediation steps that would only be taken if certain criteria were, or, were not met. The contaminant-concentration test required by Stakeholder B at Site A in the preceding paragraph is an example. There is some probability that contamination levels will exceed limits and, therefore, the remediation step would have to be enacted. Such translation of discussions with Stakeholders combined with contractor-collected data for Site A would facilitate the monetization—translation—of textual and quantitative data into a probability range that represented the likelihood that a remediation step would be necessary and costly.

. Monetization—or translation into impact—was also required of our perception of the chance that any stakeholder would prevail at any site. Stakeholder probabilities were correlated unlike the independent probabilities associated with whether or not a remediation step would be required at each site (Step A at a site might have a probability of 0–40%, while Step B at the same site might have an independent probability of 60–100%—having nothing to do with the range of probability for Step A).

For stakeholders, all probabilities at a given site had to sum to 100%. For example, if we thought that of four stakeholders at Site A, stakeholder 1 had a 60% chance of prevailing, then the sum of the remaining three stakeholder probabilities had to be 40%. This involved translating—or monetizing—our knowledge of stakeholder views and strength in a given geographic area and regarding a particular type of remediation site.

In the end, the Monte Carlo model first tested at each site to determine which stakeholder, on that iteration, prevailed. Then, the probability of each prevailing stakeholder step was tested to determine whether or not that step was required. If required, the cost range was sampled to determine the remediation-step cost. Costs were summed for each site and, ultimately, for the entire area. The major point here is that monetization involved not only "hard" data such as costs but also the translation of beliefs and political and cultural inference into impacts on the project. A successful environmental assessment resulted.

Health and Safety

No other type of project screams out more in need of risk monetization and in no other project type is monetization of risks more abhorrent to the practitioners. Typically, health and safety (H&S) efforts give a nod to the quantification and integration of risks, but due to a few important factors, true monetization of risks rarely is comprehensively embraced or implemented.

Two main impediments to employing a risk-monetization scheme in H&S projects are ethics and legalities. This statement makes ethics and legal issues sound as though they are "bad" things and, in this context, that's mainly the unspoken perception.

Ethics and the Law

Let's take ethics first. Should a company pay out more cash if a busload of elementary-school children is killed in a traffic accident, or, should it be more expensive to compensate for total fatalities in a tour bus filled with an equal number of retired persons? Are the costs associated with disability of a male field hand who was supporting a wife and five children secondary to monetary compensation paid to a disabled single female accountant? Is the loss of a leg for a mailman who walked his route worth more or less than a leg lost by a bedridden diabetic?

Making you squirm yet? And, before I receive letters regarding the paragraph above, I don't say these things in jest. Questions like these are absolutely the type with which people in the H&S risk area grapple every day.

Insurance companies have this down pat. Behind closed doors, insurance companies make these types of calculations. They do this not because they are heartless automatons, but because they are in the business of offering rational compensation for loss or injury. However, many businesses—even though insured—also have to wrestle with these quandaries. Company policies and procedures, money spent on accident avoidance, funding for safety programs, and the like are all real decisions to be made.

A glaring reality is that no company spends as much as it possibly could on H&S. For example, no company I know of sends an armored car to each employee's home in the morning to minimize the employee's probability of being injured while in transit to work. Neither do I know of any companies that provide on-site housing for all of its employees so that no employee has to travel from home to the worksite. If you truly have the safety of your employees in mind and if money is no object, why do you not provide such services?

Well, the obvious answer is that money *is* an object. Even though any company might wish to absolutely ensure that their employees would not be injured while traveling to and from work, the company can't afford to take such measures. However, the business also can't afford to do nothing at all. So, where is the balanced and practical position between absolute-assurance and do-nothing policies? Ah, this is the place at which even noninsurance company management gets neck-deep in the exercise of balancing money against injury and death. Such management would be reticent to display or discuss such calculations, but they do them.

Then there's the legal aspect. A do-nothing policy—as described in the scenario above—likely is not legally legitimate. This is to say nothing of the ethical dilemma. However, it also has been my personal experience—having been the risk expert in numerous legal actions—that no matter how much a company spends in an attempt to ward off a specific type of mishap, an attorney representing an injured party will posit that even extraordinary measures taken fell short of what should have been done. That's their job.

So, company executives find themselves positioned between ethical motivation to spend as much capital as is practical to create reasonable assurance that a specific foible will not materialize while knowing full well that no matter how much they spend, if the mishap comes to pass, their efforts and expenditures might be found to be lacking. Monetization of the circumstance and attendant risks can aid them in finding a capital spends that is rational and reasonable.

Monetizing H&S Risks

As described in some detail in Chapter 3 (in the section "Uncertainty and the Confusion that Surrounds It"), a measure of frequency is a view of the past. In the H&S world, much of the statistical information to be found relates to frequency,

For example, it is common to find charts and graphs which plot things like population density against number of accidents of one sort or another. Age-of-driver vs. accident severity and other such plots are ubiquitous. These are all counts of things that have happened. However, it is my experience that such counts are rarely the best harbingers.

I once again fall back—as I did in Chapter 3—on the coin-flip scenario. Just because I flipped the coin twice and got two "tails" has absolutely no bearing on the probability of getting a "tails" on the next flip. The probability of getting a "tails" is 50%, no matter what is the count of tails I have gotten in a fixed number of flips in the past.

Frequencies are great pieces of information, but it might not serve as the best cornerstone of any estimation of the probability of the event happening in the future. Just for example, I might be counting the number of accidents at a city corner over the past 5 years. The number is, say, 100. So, I could deduce from this frequency information that in the coming year, I can expect about 20 accidents. This type of count, among other things, ignores trends in the data. There might have been an inordinate number of accidents at that corner until 3 years ago when a traffic light was installed a half mile down the road. Timing of the changing of the new traffic light was coordinated with the changing of the traffic light at the corner in question so that traffic now flows much more smoothly. As a result, there have been but six accidents at the corner in question over the past 2 years. Ignoring trends within counts is done at one's own peril.

Trends can work both ways, however. If I resort once more to the coin flip example, I might have had exactly alternating "heads" and "tails" for 20 flips and then had a trend of five "heads" in a row in the last five flips. Should I take this trend into consideration when calculating the outcome of the next flip? Of course not. So, paying attention to trends within counts is done at one's own peril.

Well, then, just what should one do to generate probability-of-future-occurrence data? I'll bet you can guess what I'm going to tell you.

H&S Probabilities and Black Swans

First, let me attempt to address the "black swan" issue. This subject was eloquently addressed by Nassim Nicholas Talev in his book *The Swan—The Impact of the Highly Improbable* (see the section "Selected Readings" at the end of this chapter).

No matter how many risk-identification workshops are held or what level of diligence is employed to scan the horizon for new risks, it is inevitable that at least one significant threat or opportunity—whether it materializes or not—might be overlooked. This is where probability and black swans get into a nasty tangle.

When you are working with a portfolio of projects, the intermingling of probabilities and, say, project costs is clearer. For example, imagine you had a portfolio of 10 projects to potentially be executed in 1 year. Each project, if executed, will cost $100. In addition, each project has a 50% probability of not being executed in the coming year. It is pretty clear that if you budgeted $500 for the year, you might be close to the actual cost of the executed projects. Of course, in any given year you might spend more or less than the probability-weighted amount, but over a number of years, your total spend should come close to the sum of the probability-weighted values.

Not so with a single issue such as a black swan. You might be astute enough to realize that something "big and bad" might happen, but that you just don't know what it is. Black swans are black swans because they don't happen often (profound, huh?). That is, the probability of the big, bad, and expensive event is exceedingly low. If you set aside the probability-weighted amount of cash, that amount would be of relatively small magnitude—a tiny probability times a large amount of cash results in a number that is "small" relative to the actual amount needed if the event comes to pass.

Even if you have a portfolio of black swans and their attendant costs (you poor thing!), the cash set aside for the multiple events—because of the exceedingly small probabilities—likely will fall far short of what is required. The only way this works out is if you set aside a new pile of risk-weighted money every year and add it to all that was set aside from previous years. In that way, the cash reserve has some chance of reaching a level that might actually be useful if the black swan hits. However, this is not how businesses typically operate. Even if businesses acted in such a manner, just like pension funds, it is incredibly tempting to tap that cash for other purposes.

So, the bottom line is that if you believe that you can be bitten by a black swan, you have to either accept the fact that you will not have sufficient funds set aside when the swan bites (risk acceptance), or, you will set aside an amount of cash that squares with your perception of the cost and just let the cash "sit there" in anticipation of the exceedingly unlikely event. As I said, many H&S events are of this exceedingly rare type—explosions, mass deaths, and the like.

Putting aside the black swan and assuming that one should not rely only on past frequency-of-occurrence data, just how should anyone estimate a probability for a H&S event? The first part of this answer is to define the threat in the recommended manner such as:

> If the chain on the rotating machinery breaks (Reason), then dismemberment of a worker or the demise of that same worker is a possibility (Event) causing the company to pay all associated medical and other costs (Impact).

Now, I have written this threat in a relatively general manner, but any real-world risk should include as much detail as is practical. Right away, however, we run into trouble.

Any attorney worth her or his paycheck would advise the company that issues that paycheck not to record risks in this manner. Such documentation of the threat can be tantamount to admitting that the company knew about the potential problem and, as any good prosecuting attorney would argue, the company didn't do enough to prevent the event. This scenario regarding the documentation of H&S risks, and therefore the lack of monetization of those risks, should be expected to be a hurdle.

Providing such legal intervention did not occur, or, was ineffective, the probability of the risk described should be determined in the same manner in which any risk probability is estimated—by conversations with discipline experts. Just like the gathering of experts at a risk-identification workshop, a meeting of knowledgeable individuals should be convened. At that meeting, any relevant frequency data should be available for perusal and discussion. Experts, however, should know how to temper the frequency information to arrive at a consensus range of probabilities for the event. See the "teams" story in the section "Uncertainty and the Confusion that Surrounds It" of Chapter 3.

For example, in the aforementioned accident-at-a-corner scenario, experts should know that an average of the 5 years of accident data should not be used as a projection because of the installation of the second traffic light 3 years ago. Experts would also know that any trend of five "tails" in a coin-flip situation has nothing to do with the probability of the next flip. A panel of experts might not be able to come to consensus on a single (deterministic) probability value, but an experienced facilitator can nearly always glean from such experts an agreed upon range of probabilities. That distribution of likelihoods can be used in the Monte Carlo risk-monetization process.

Consequence Ranges for H&S Projects

Legal impediments are a problem regarding establishment of probabilities for H&S risks. However, those hurdles pale in comparison to the angst you will experience when attempting to define consequence values or ranges for H&S threats and opportunities.

Turns out, there's no joy in assigning cash value to sensitive things like loss of a body part, physical and mental disabilities, or death. It's bad enough when a value has to be put on, say, loss of an arm. It is at least doubly difficult to establish the relative value of two different arms.

Is, say, the arm of a low-wage worker worth more or less than the throwing arm of a professional baseball pitcher? How about the leg of a professional dancer relative to that of an accountant? What about their relative ages? Are "young" legs worth more than "old" ones? Sensitivity enters the picture when one attempts to assign a value to an attribute of a specific person or cohort of people.

Of course, one way to slay this dragon is to just face the music and make assignments of value. Cold, I know, but necessary. However, if your task is to ascertain the probable costs of, say, limb loss for a cross section of workers—workers who individually would belong to multiple cohorts with regard to age, social status, etc.—then the task is a bit more palatable.

Suppose you are assigned the unenviable job of determining the liability costs associated with car-accident injuries suffered by company employees who drive as part of their job. The population for which you are making estimates ranges from top executives who drive to various company sites for meetings to low-wage field hands who monitor equipment in remote geographic areas.

Certainly, severity of injury would have to be considered. Employees of relatively high rank might be more likely to employ private attorneys and so might present a greater threat of litigation. Therefore, you might take the attitude that injury to a top executive could be more costly than an equivalent injury to a company worker on a lower tier of the organization. Behind closed doors then, you might establish a distribution for the injury type in question. The low-cost end of the distribution might reflect the lower values related to salary-adjusted costs, whereas the upper end of the distribution would represent the impact of higher salaries translated by an algorithm into compensation cost.

When presenting the distribution, it need not be explicitly revealed that one end of the range represents one "type" of worker and the opposite end typifies an employee of a disparate sort. All that need be revealed is that the range represents the reasonable spread of costs associated with this type of injury and severity of injury for the company population. This, typically, is a more palatable means of monetizing a risk.

Threats and opportunities and their associated probabilities—all discussed above—need to be monetized and integrated with similar and other data if a true—or as close to true as is practical to hope for—estimate of impacts from H&S issues is to be had. Later in this book, I will give specific examples of such data integration—though not in an H&S setting. The bottom line is that you should be aware that risk monetization is absolutely essential in the realm of H&S risk assessment and management, but it is also the arena in which legal and other forces combine to make it one of the most difficult areas in which to enact a proper risk assessment, risk monetization, and risk management effort.

Risk Monetization of Legal Matters

I don't mind at all admitting here in writing that I was absolutely wrong about something. Prior to becoming involved in the risk analysis of, literally, dozens of legal matters, I had a pretty dim view of the folks who make up the legal profession. I believe it is part of our US culture to need lawyers, but to not be very fond of them. I pretty much harbored such an attitude until I became involved in the risk analysis of complex legal issues. Not only did my attitude change for the better, but I gained a newfound and genuine respect and admiration for at least a segment of the population of attorneys.

First of all, most of the legal eagles with whom I have had the privilege to rub elbows, so to speak, have been really smart folks. Certainly they exhibited native intelligence, but they were also astute enough to recognize when the legal matter at hand would benefit from a proper assessment of risk. The number and type of legal cases upon which I have served as risk expert are great and the spectrum of legal issues is equally broad. I will here relate just a few of the risk assessment or monetization efforts so that the reader can realize that risk analysis has a place in legal matters.

Estimating Potential Liability

I have never been much of a "company guy." I have had associations with many coworkers who, if you cut them, would bleed the company colors. These were folks who had a real and tangible loyalty to the company and who believed the company returned the sentiment. That was never me. I always viewed my employment by a company as a business deal. I agree to show up and do what I'm supposed to do and they agree to pay me. Every payday, we "square up" and start again. If I fail to show up, they can stop paying me and if they stop paying me, I'll stop showing up. It is simple and cold as that.

In spite of my business-like view of my relationship with any company, every once in a while an organization would surprise a person like me a bit with regard to their truly ethical or "high-road" approach to a situation. Consider the following story as an example of such behavior by a company.

First, company attorneys will approach you and suggest that you sign all sorts of documents—most stating that you'd better not breathe a word concerning what you are about to learn. Subsequently, they revealed that, truly, through no fault of its own, some company equipment had been damaged (by a third party) and that company materials had been released into the world. Those materials caused damage of various sorts. Enough said about that.

Their question to you—handed down to them from "on high" in the company—was not one relating to how the company might "beat the rap" or how they might minimize their perceived liability or any other such potential attempt to minimize reputation, actual, or punitive damages. Rather, what they ask you to do is to take a cold and impartial look at the case and give an estimate to the company regarding how much "cash" they should set aside—on a probability-weighted basis—to adequately settle the case and to justly compensate the alleged injured parties should it be decided that they needed to do so.

You might find this attitude especially refreshing because you are just the guy they would come to if they wanted to estimate their chance of prevailing with any given legal strategy or to design a risk-based scenario in which they would attempt to minimize their culpability. They don't do that. They simply want to know, no matter how the case was ultimately decided (favorable to them or otherwise), just how much they should expect this to cost on a probabilistic basis.

To relate all aspects of such a case would be to write another entire book. It can become very complicated very quickly. In the end, such a case can involve more than a dozen companies, individual attorneys representing perceived injured parties, insurance companies, and municipalities.

As with most legal cases that are heard by a judge or a judge and jury, venue is important. Where a case is heard is nearly as important as the arguments made in court. Therefore, one of the first parameters to be evaluated and monetized is the venue.

This is accomplished by engaging in extensive conversations with several defense attorneys. Through facilitated conversations, their views on biases, track records in front of local judges, local news media, president, local laws, and other considerations are captured for all potential venues. This information is translated—by the attorneys—into a range of chance-of-success weighting factor for each potential trial location.

Quality of council also is a critical factor. Defense attorneys and their staffs can review the trial records of the firms and individual lawyers that likely would constitute the prosecution. As with the venue translation described above, through facilitated conversation, you prompt the defense council to estimate a chance of success associated with each potential prosecuting law firm or individual prosecuting attorney. Again, like the venue, this information was utilized as a weighting factor in the Monte Carlo model.

Other more mundane but equally crucial information should be collected regarding potential fines, possible real and punitive damages, attorney costs, and many other possible negative impacts. Of course, you should also consider the possibility of prevailing. However, because there would be no monetary prize associated with a "win" and because winning is not free, the primary output metric from a computer program—typically, thousands of lines of code—was cost. Of course, separate ranges for associated probabilities need to be generated for most of the aforementioned parameters.

Modeling this situation can be a programming nightmare mainly because of all the possible permutations. Just for example, logic needs to be constructed for scenarios like: "If company A is granted a positive summary judgment, then companies B and C will likely prevail causing municipality D to ..." With more than a dozen entities in the fray, the amount of recursion in the model can be great. In the end, though, all major scenarios can be accounted for and a probabilistic estimate of damages generated.

My point in relating this story is to impress upon the reader that monetizing risks—turning threats and opportunities into impact on value—is not only relevant, but essential when evaluating complex or even simple legal matters. Translation of parameters such as venue, quality of council, precedence, and other "soft" risks (as opposed to threats such as costs) is necessary, possible, practical, and exceedingly useful.

This example is one of significant complexity. However, monetization of threats and opportunities applies equally to relatively simple legal matters, as I will relate in the following section.

Decision Trees Applied to Legal Matters

As a means of calculating probabilistic outcomes, I am not nor have I ever been a big fan of decision trees. In real life, things are more complicated than the more-or-less linear logic the tree can easily represent. The "If/Then/Else/While" type loops, time-series analysis, recursion, and other advantages of computer programs generally make those programs superior vehicles for emulating reality. In addition, in complex cases, it is not long before a decision tree morphs into a decision bush [see page 145 of *Risk Assessment and Decision Making in Business and Industry—A Practical Guide*, 2nd edition, by Koller (2005)]. That is, the addition of just a few new parameters to a decision tree can result in an explosion in the number of leaf nodes.

Having said that, however, I and many others have found that decision trees are excellent vehicles for discussing or "talking through" a problem. Trees facilitate the conversations and "force" people to think about what variables pertain, the likelihood that scenarios (branches of the tree) will materialize in real life, the magnitude of the coefficients associated with outcomes (leaf nodes), and the sequence in which things might happen in the real world. So, although I would not choose a decision tree as a calculation vehicle, I find them exceedingly useful when attempting to glean from clients the essential pieces of information and how those items are temporally and logically related.

I also have always been a fan of solving decision trees backward relative to the way they traditionally are solved. Typical solution methodology is shown in Figure 4.1.

In the top "Tree A" shown in Figure 4.1, the values at the leaf nodes are deterministic—that is, they are single values. The probabilities associated with each of the branches (40% and 60%) are used as multipliers. That is, each leaf-node value is multiplied by its associated probability and a "weighted average" or "expected value" is generated.

Replacing the deterministic leaf-node values with ranges (distributions) really does not change things much. In the lower "Tree B" in Figure 4.1, the tree was solved 1,000 times. On each solution, a value was randomly selected from each leaf-node range and that value was multiplied by the branch

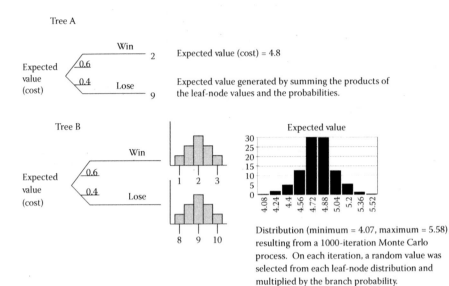

FIGURE 4.1
Two simple legal-analysis (win or lose) decision trees—one with single values (Tree A) at the leaf nodes and one with ranges at the two leaf nodes (Tree B). Both are solved using probabilities as multipliers.

probability. The two resultant products were summed to arrive—1,000 times—at an expected value. The range for the expected values is shown on the left side of Tree B.

Resultant expected values in Tree B are great if what you are after is a weighted average, or, a range for the weighted average. However, in real life, what people really want to know is: "How bad could it be, how good could it be, and what is the likelihood that it will be good or bad?"

For example, even in the severely oversimplified legal scenario outlined in Tree B of Figure 4.1, anyone subject to the outcome of the legal case is going to want to know just how bad things might be, or, if you really expect to win, just how little it might cost to litigate. To answer such questions, it is much more telling to use the branch probabilities as "traffic cops" rather than as multipliers.

To use probabilities as "traffic cops," the algorithm starts at the left end of the tree. A random number between 0 and 1 (or, between 0 and 100) is generated. If, as in the tree shown in Figure 4.2, the random value is between 0% and 40%, then the top branch is taken and a random value is selected from the leaf-node distribution and is saved as a result on the outcome distribution shown at the left end of the tree. If, however, the random value is between 40% and 100%, then the bottom branch is taken and a randomly selected value from the lower branch leaf-node distribution is selected and

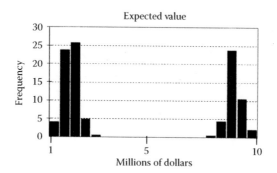

FIGURE 4.2

Result from using probabilities in Tree B in Figure 4.1 as "traffic cops." The two sets of bars shown here are the leaf-node distributions shown on Tree B in Figure 4.1 with the frequencies impacted by the probabilities (.6 and .4). Note that the Expected Value range shown for Tree B in Figure 4.1 is not a possibility in this plot.

is saved as a result on the resulting outcome distribution at the left end of the tree.

Regardless of the number of branches or leaf nodes in the tree, if the branch probabilities are used as "traffic cops" (random probability value generated at each branch and the corresponding branch taken), the result will be the selection of just one value from a given leaf-node distribution for each solution of the entire tree.

In Figure 4.2, the distribution shown at the left end of the tree is not a range of expected values (although, the mean of this distribution is precisely the same as that of the distribution shown at the left end of Tree B in Figure 4.1—the expected value is not lost). This distribution has a much broader range than that of the expected value distribution of Tree B in Figure 4.1 and relates to the observer the range of possible outcomes and the probabilities associated with those potential results. This is what most folks really want to see.

Now, real-life legal cases typically are more complicated than the simple representations in Figures 4.1 and 4.2. In Figure 4.3 is shown a still simple, but more representative scenario.

In Figure 4.3, I represent a straightforward legal strategy. Initial efforts will be made to settle the case out of court. If the legal team is unsuccessful at settlement, then the next step is to work to win a summary judgment. Winning the summary judgment does not result in a benefit and the value at the end of the "win" branch represents the cost of a "win." If the summary judgment is not in favor of the legal team, then either actual damages, or, actual damages plus punitive damages will be assessed.

Plot A at the bottom of Figure 4.3 shows the expected value range resulting from the solution of the decision tree in the traditional way—that is, by using the branch probabilities as multipliers. Plot B in the same figure shows the range of probable outcomes resulting from solution of the tree utilizing

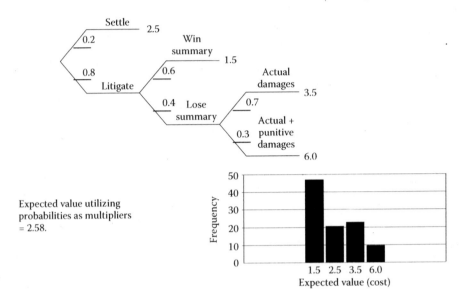

Expected value utilizing
probabilities as multipliers
= 2.58.

Frequency and range of expected
values when probabilities are used
as "traffic cops." Use of distributions
at leaf nodes would create intermediate
values on the X-axis.

FIGURE 4.3
More complex (than that shown in Figure 4.1) legal analysis contrasting the result generated by
using probabilities as "multipliers" versus using probabilities as "traffic cops."

branch probabilities as "traffic cops." Plot B indicates the range of possible
outcomes and associated likelihoods. The "expected value" (mean) for both
distributions—Plot A and Plot B—are identical.

Other Project Types

In this chapter, I have made example of just a few project types (acquisitions,
divestitures, health and safety, environmental, and legal). Hopefully, I have
made the case that risk monetization can be—and should be—applied to
the ilk of efforts delineated above. However, the types of projects to which
risk-monetization techniques can be successfully and practically applied are
nearly endless but would certainly include the following:

- Portfolio analysis
- Construction projects

- Decommissioning
- Equity positions
- Production forecasting
- Employee services

And many other project types. Readers should feel free to contact me for help with any project type.

Suggested Readings

In addition to books and articles cited in the preceding text, this list includes other material that the reader might find helpful and relevant to the subject matter discussed in this section of the book.

Bartlett, J., et al. *Project Risk Analysis and Management Guide.* 2nd ed. Buckinghamshire, UK: APM Publishing Limited, 2004.

Bayes, T. An Essay Towards Solving Problem in the Doctrine of Chances. *Philosophical Transactions of Royal Society* 52, 370–418, 1763.

Koller, G. R. *Risk Assessment and Decision Making in Business and Industry: A Practical Guide.* 2nd ed. Boca Raton, FL: Chapman & Hall/CRC Press, 2005.

Taleb, N. N. *The Black Swan—The Impact of the Highly Improbable.* New York: Random House, 2007.

5

Without Risk Monetization, What Typically Is Done

To appreciate the impact of risk monetization, a baseline must be established. That is, risk monetization is a process that can improve project performance relative to such performance sans monetization. So, if I am to relate to you the betterment of projects due to application of risk monetization techniques, I first have to establish just what is now typically done with regard to risk assessment and management.

Many of the points I will make in this chapter have been addressed at some level in the previous sections of this book. Therefore, rather than dwell on certain aspects again, I will simply refer to previous treatment. However, in this chapter, I wish to coalesce and consolidate previously mentioned and new elements of the typically employed risk process. In Chapter 6, I will describe in detail remedies for the various shortcomings documented here.

Project Envisioned

In the section "Be Clear as to 'What Is the Question,'" of Chapter 3, I establish that representatives from different disciplines can harbor unique views of a given project. It is typical, for example, for a project manager, a construction engineer, an environmental technician, a negotiator, a commercial analyst, a financial analyst, a legal consultant, and others to all envision unique concepts regarding project starting dates, project scope, what tasks are most important, project timelines, and the like.

When discussing the project and when identifying risks—such as in a risk-identification workshop (RIW)—it is critical that all parties are talking about the same project and not different perceptions of the project. For example, in a RIW, it would not be out of character for a negotiator to envision the project beginning with initial contacts with the host foreign government and ending with successful negotiation of a mutually agreeable contract for project initiation. I am not here intimating that such discipline experts are short sighted, but it can be difficult to impossible for such representatives to imagine how early-in-the-project decisions might impact, for example, operations at the end of the project lifecycle.

It has been my experience that at initial gatherings of discipline experts, the sign outside the meeting-room door might say "Project X Meeting" and everyone inside the room would swear that they are there to discuss Project X, it is nonetheless true that there are as many visions of what constitutes Project X as there are people in the room. To not nip this significant foible in the bud is a huge mistake. In Chapter 6, I will give an example of a remedy.

Incomplete Representation

This issue typically is the one that establishes the power hierarchy in the risk assessment or management process. For any project of significant size or importance, a person with a title such as "Risk-Process Proponent" (RPP) should be identified. This individual is charged with responsibility to shepherd the project group through the risk process and to facilitate the successful execution of each risk-process-related step.

A first test of the power vested in the RPP usually is the convening of an initial RIW. Often, the RIW is a multiday affair demanding several days of commitment from each participant. It is tough enough to schedule such a gathering such that it does not conflict with already established commitments for one or more potential participants. It is an equally arduous task to gain commitment from the identified representative from each critical discipline.

A construction project, just for example, might demand participation by representatives from any of the following:

Union

Negotiating team

Health and safety department

Logistics group

Environmental affairs group

Legal department

Construction engineering department

Finance group

Commercial analysis group

Design department

Project management group

Process engineering department

Others

A RPP must be someone (or a small group of people) who is attuned to the workings of the greater organization. This is true for at least two reasons. The RPP must be astute enough to realize which disciplines need to send participants to the RIW. Secondly, the RPP must know which individuals within each required department can best represent the specific discipline. If a critical discipline is absent from the RIW, or, if an incompetent representative participates, the RIW will fail to achieve its goals.

When the RPP issues a request to each critical department for specific representation, one of four things can happen. One reaction is that department heads respond to the RPP request for participation with a refusal to commit already busy personnel to a multiday RIW. A second response is to identify someone in the department that has little else to do and to assign that woefully unprepared person to be the department's RIW attendee. The third and most insidious tactic is to initially agree to send the requested representative and then, at the last minute, send instead an excuse why said representative can't make the meeting. The fourth scenario is one in which the discipline departments earnestly comply.

In the preceding paragraph, the first three described scenarios constitute noncompliance situations. Such situations typically represent the first test of wills and power. The RPP should refuse what she or he decides is unacceptable representation. Agents from each discipline should not only understand all aspects of the specific discipline, but should have a working knowledge of how their discipline interacts with other disciplines in the project setting. Refusal by the RPP to accept unqualified representation from the discipline almost always precipitates involvement of management levels above the RPP. If the RPP's management does not stand behind the RPP's request for a competent emissary, then the risk process is in deep trouble right from the start.

Refusal by the RPP to go forward with a poorly attended RIW also can set up clashes. For example, a RPP might have initially gotten a commitment from each discipline regarding requested representation at the RIW. However, on the day of the meeting, critical individuals are suddenly "too busy" with unforeseen commitments and can't make the RIW. At this point, the RPP should not proceed with the RIW and should attempt to reschedule the meeting as soon as possible. This approach, of course, will not please project management who are committed to a project schedule and such management will not appreciate delays.

Again, the management of the RPP is likely to get involved and those managers need to support the RPP's decision to cancel the RIW and to reschedule. If the RPP managers stand their ground, discipline representation at a subsequently scheduled RIW will be much improved mainly because the project manager will bring significant pressure to not have a second failed RIW. Attitudes of the representatives, however, might reflect little improvement.

Don't Tell Me What to Do!

If the preceding paragraphs intimate that I have personally been down this road, then the intimation has been successful. Although most managers and their reports will "talk a good game" when it comes to risk assessment and risk management, those same reports and their leadership usually just want to get on with their work and do not appreciate several aspects of the risk process.

A first aspect of the risk process for which project team members lack appreciation is that they don't like "being handled" by an "outside" entity—namely the RPP. On most project teams, the RPP is held responsible for conducting a succinct and comprehensive risk analysis of the project and for overseeing the management process for the identified threats and opportunities. This puts the RPP squarely in the line of accountability which, of course, becomes an issue if a threat, for example, materializes—especially one that had been previously identified.

A second aspect of the risk process that generally generates push-back by project team members is the interdisciplinary attribute. For example, a design team might consult with construction engineers when designing major hardware for a project. However, those same team members rarely can envision why they might need to discuss their part of the project with legal staff, representatives from logistics, negotiators, environmental experts, and other seemingly unrelated areas of expertise. Utilizing the design staff just as an example, I can't tell you the number of times that the design of major hardware (or software) was found—sometimes after it was actually built—to be inadequate for the job. Legal restrictions or local laws, environmental constraints, transport limitations, original negotiated-contract constraints, and a host of other considerations should have impacted design, but did not.

It is just a fact of business nature that any given group is comfortable dealing and exchanging information with a small cadre of other disciplines. Getting those groups to recognize the holistic nature of a project and the interdisciplinary nature of the entire effort is a tall task. The RIW event is one of the few—if not the only—forums in a project life when seemingly disparate discipline representatives will cross paths. It is the job and responsibility of the RPP to convey to all representatives the criticality of a holistic approach to threat and opportunity identification and a community approach to managing the risks. I will relate much more in later chapters on just how this might be accomplished.

Poor Translation of Qualitative Data into Quantitative Data

Although the mantra of this book is to monetize threats and opportunities, it is a rare instance when all risks—especially those based on qualitative data—are monetized in a meaningful manner. In a RIW, for example, it is common to witness the expression of the impact and/or probability associated with some risks in a mainly qualitative way. While it is true that in the section "Monetization" of Chapter 3, I state that consideration of threats and opportunities in a qualitative manner is an acceptable monetization technique providing the threats and opportunities so considered impact the perceived value of the project, such qualitative treatment is not the preferred route.

It is a relatively trivial endeavor to express more "traditional" risks with quantitative metrics. For example, if a risk is strictly a cost, then it is no great stretch to have the range for the risk impact expressed in dollars (or any other currency). Similarly, if an opportunity arises that will advance the schedule, quantitative representation of the advancement is the norm—whether expressed in time units (days, weeks, etc.) or in monetary impact.

Three primary impediments to qualitative-risk translation are lurking out there. One is the general inability or willingness of project personnel to deal with "soft" risks. Another is the format in which some risks are expressed. A third is the lack of a formalized and structured method for meaningful translation of risks from a qualitative to a quantitative format. I'll here address each of these issues.

Ignoring Soft Risks

As I have reiterated numerous times in this book, soft risks are those that typically are not line items in a model that integrates "all" aspects of the project (usually this is a spreadsheet-based financial model). Not only do such risks lack specific representation in the model, but addressing such risks can be a very uncomfortable exercise for project personnel. I'll here offer two examples.

A more traditional "soft" risk is fatalities. It is not unusual to see in a project model a line item that addresses health and safety costs. However, it is exceedingly rare to see a specific entry that expresses the impact on the project emanating from fatalities. Especially in the U.S., the basis for avoidance of such data from the model is the litigious environment. Corporate attorneys rarely sanction the expression in a model of the possible number of fatalities or their impact. If there is any consideration at all of such a risk, it usually is "hidden" in other parameters such as "contingency" or some other catch-all element. There are typically three ways of dealing with this.

First, risks such as fatalities can be addressed in a separate and "sealed" risk assessment that is passed directly from the legal staff to the top decision makers. In this case, probabilities and impacts of the risks are not technically integrated with other risks but are conceptually considered by top brass. A second method of dealing with this type of risk is to develop a culture in which such risks are among those that are—without specific documentation—accepted as being omnipresent. In such cultures, however, it can be difficult to, for example, mount a campaign to reduce fatalities because the subject is avoided in any analyses or interactions. A third method—alluded to in a previous paragraph—is to "hide" the impact of such "soft" risks in the consequences of other less controversial considerations. For example, fatalities might not be specifically mentioned in the assessment, but health and safety costs might be traditionally inflated by a "fudge factor" in an attempt to surreptitiously account for the impact of potential fatalities. Of the three processes mentioned here, I personally favor the first—that of creating a proper risk analysis for the "soft" risk but to safeguard the information, to restrict access to the assessment, and to eschew integration of such information with the bulk of the risk data.

Other types of soft risks that can be a source of great discomfort are those, just for example, that address cultural issues. It might be that a project is being pursued in a country in which religious traditions will significantly impact efficiency of execution. This might stem from outright bans on certain activities that are perfectly acceptable in the country initiating the project to gender issues to concerns about process interruptions for traditional prayer. It is a rare instance when a project model specifically addresses such issues, however, those issues, when not illuminated and considered, can lead to at least real cultural clashes and at worst a significant adverse event. As uncomfortable as it might be, such potential risks have to be addressed and specifically illuminated and enumerated in the project model. Political, organizational, and other "soft" risks must similarly be recognized and addressed.

Format for Risk Identification—Example of the Peer Review

I mentioned above that three primary impediments to qualitative-risk translation are lurking out there. One is the general inability or willingness of project personnel to deal with "soft" risks and this aspect I have addressed in the immediately preceding paragraphs. A second source of qualitative-risk translation is the format in which some risks are identified. This is best illustrated by the peer review.

Let me state from the start that I am not a big fan of the peer review. I do agree with the intent. Peer reviews traditionally are envisioned as formats in which disinterested but knowledgeable parties meet to discuss and "pass judgment" on the project of a peer. If in reality such reviews worked as intended, then all of my objections to this format would be moot.

However, I have rarely witnessed such a review that actually operated in an unbiased manner. The "good old boys" club atmosphere and especially the "I won't hurt you in reviewing your project if you won't hurt me when my project is reviewed" attitude—and many other things—tend to make peer reviews less effective than one might hope they would be. However, the effectiveness of this format is not really the issue here.

I am all for discussion and conversation regarding a risk. If addressing risks goes only as far as conversation in a peer-group review, however, then integration of those risks and illumination of their impact on the project is, at best, exceedingly difficult. Peer group reviews tend to be attended by relatively senior staff members. Most attendees, therefore, do not have as part of their job description the taking of notes or the translation of any notes into quantitative format. Such meetings commonly conclude with lists of recommendations and/or actions to be taken but rarely with a relative ranking of the impact of those actions on the success of the project. Such prioritization is necessary if significant time and effort are not to be wasted in the pursuit of relatively ineffective measures. Such prioritization is the crux of the monetization process.

Translation of Risks

Because of the intricately intertwined nature of all of the "pieces" of the risk process, in writing this book I am regularly in danger of getting ahead of myself. This is one of those instances. This section addresses the third impediment to qualitative risk translation.

In Chapter 6, I will take up the issue of utilizing a table to translate—at least initially—risk consequences into impact on value. However, because attempts at translation of risk consequence into value either typically are poorly performed or not done at all, it is appropriate to address the issue here. This section also relates to the section "One, Two, Three—GO!" later in this chapter.

In a RIW, it is common to collect information and data regarding the probability and consequence of each identified threat and opportunity. When metrics align, it is typical to attempt to rank risks by generating a product from the probability and consequence—that is, multiply them. Sometimes this works as an initial attempt at ranking. If risk-consequence data have all been translated into, say, dollars, then this approach is possible. However, it is not difficult to imagine that one risk's consequence is expressed in dollars while schedule slip (metrics of days, weeks, etc.) is used to express the impact of another risk. In such cases, comparing for ranking purposes, the product of probability and time with that of probability and dollars makes no sense. A solution to this problem will be presented in subsequent sections of this book.

Another tactic universally employed to attempt to initially rank risks is that of setting up categories. For example, a project team might establish three categories for dollar impacts and three categories for schedule impacts. For threats, team might use the following descriptions:

$0 to $500,000 impact = LOW
$500,001 to $1,000,000 = MEDIUM
>$1,000,000 = HIGH
0 weeks to 2 weeks delay = LOW
2 weeks to 2 months = MEDIUM
>2 months = HIGH

Then, typically in a spreadsheet, the consequence of each risk is compared to the established categorical ranges and assigned a LOW, MEDIUM, or HIGH label. In this way, risks expressed in any number of metrics (time, money, etc.) can be initially compared and ranked.

Yet a third common mechanism for initially ranking risks is to use a more sophisticated translation table. For example, the commercial analyst might be queried regarding how much impact on Net Present Value (NPV) (or any other financial metric such as IRR, DROI, etc.) results from various risk impacts at 100% probability. See the section "Create Translation Table" of Chapter 6 to see how risk probabilities and impacts affect the translation-table process.

The analyst might be asked to translate into NPV a cost of between $0 and $500,000, between $500,000 and $1,000,000, and so on. Similarly, the analyst might also be asked to run her or his financial model to determine the NPV impact of up to a 2-week delay, a 2-week to a 2-month delay, and so on. This method facilitates an initial ranking of risks by a common metric (NPV or whatever) and utilizes the actual financial model for the project.

However, even this process falls short of actual monetization of the risks. Correlations between threats or opportunities, the time value of money (all consequences do not impact the same time-series segments), and other con-siderations make even this method a poor second to formal monetization of risks. The section below—"One, Two, Three—GO!"—expands on this concept.

One, Two, Three—Go!

In the previous section, I opened the door to the world of risk ranking. In that section, I focused on some of the methods typically utilized to attempt to initially translate risk impacts into some sort of metric that can be used to rank threats and opportunities. In this section, I will concentrate on the issues surrounding ranking of risks regardless of the method employed to translate risk probabilities and impacts.

In the process of ranking risks, problems can arise that are similar to those realized in the game of portfolio optimization. Imagine that you have four departments in your company—departments A, B, C, and D. Further con-sider that each department has multiple projects that it would like to have funded in the coming fiscal year. The company can, however, only afford to fund a fraction of the total number of projects in all departments. You have given the head of each department instructions to prioritize and rank the potential projects in their respective areas.

Faithfully, this is done. Department A has five top-ranked projects (A1, A2, A3, A4, and A5) with project A1 being their most "important" potential effort.

Department B has seven projects in their most important list, Department C has four, and Department D has nine.

One way, but a flawed way, to create a corporate portfolio of most important projects is to take the top N projects from each area. That is, the top two projects from department A, the top two projects from department B, and so on. Let's say we are using potential after-tax revenue as the project-ranking metric. The glaring foible in this approach, of course, is that it could be that the "best" project from department A (project A1) has significantly lower revenue-generating potential than the worst projects from all other departments.

Clearly, then, if we include any of department A's projects in our final list of projects to fund, the portfolio will not be one of optimal potential company revenue. Of course, in a real-world portfolio—assuming department A is a strategic element of the company—some of department A's projects would be included because, otherwise, most or all of the employees in department A would have nothing to do—and you know what that means.

So, the technically optimum portfolio—that is, the portfolio with the best revenue-generating potential—would not include any projects from department A. This analogy can be applied directly to the use of risk registers to rank risks.

Suppose we have the same four departments—A, B, C, and D—that have each listed some number of risks in the risk register. Each risk has an associated probability, impact, capture, or mitigation action, and the cost for that action. Especially in the Web-based situation in which each department has appointed a unique individual to tend to that department's risks, there is the tremendous urge for each department to rank their risks and set about mitigating those that bubble to the top of their list.

Just like the portfolio of projects, the company can afford to address a relatively small subset of risks. That is, the opportunity-capture or threat-mitigation costs represent real capital outlay for the coming fiscal year and the company must address only that subset of overall risks that are most important.

Determining which risks are "most important" might be tricky. Some companies wish to use the unadulterated impact of the risk as the ranking mechanism. These might be companies in which they have some risks with associated tiny probabilities but relatively huge impacts (the 747 falling on the plant, for example) and they don't want to make the risk "disappear" by multiplying the huge impact by the tiny probability. Other organizations tend toward using the probability as a multiplier for the impact so that a "weighted" impact is the ranking criteria. This is a useful approach when the probabilities associated with risks do not tend toward the infinitesimally small. Organizations in which all impacts are about the same magnitude (e.g., a company in which, say, the digging up of buried tanks at hundreds of different sites would at each site cost about the same amount) might opt to use probability (of tank failure) alone as the ranking mechanism. Still other organizations might include the cost of capture or mitigation in their ranking calculations (not a good idea—more on this later).

Regardless of what metric is used to rank the risks in the limited-budget scenario, what we likely don't want is each department ranking their risks and then (one, two, three, go!) setting out to capture or mitigate their top risks. Just like in the portfolio example, it might be that because of their low ranking relative to the risks in other departments, none of the risks from department A should cause us to spend capture or mitigation capital. However, when you employ a Web-based risk register that can be simultaneously accessed by different departments, it is a bit of a task to prevent the "one, two, three—go!" situation from happening. It is even more of a challenge when each department keeps its own risk register or, no register at all.

Again, after all risks are initially entered into the risk register, a whole-register ranking of the risks should be undertaken by the RPP. This is the only practical way to attempt to ensure that money is not wasted on risks that should not be addressed—at least not yet.

The last thing I will mention here is something I alluded to above. It is generally, but not always, a poor idea to include the cost of risk capture or mitigation in any metric that will be used as the basis for risk ranking.

One of the decisions to be taken regarding any risk is "Is it worth it?" That is, regardless of whether you utilize the probability as an impact modifier or not (multiply the impact by the probability), one of the decisions to then be made is whether or not the magnitude of the impact justifies the cost of the capture or mitigation action. Now, those of you who are astute in the ways of risk monetization might realize that one of the aims of risk monetization is to increase project *value* and considering costs alone should not be the aim (e.g., it might yield high value to spend lots of money). While this is true for the overall project, when considering individual risks in the ranking process it might be OK to consider a simple cost–benefit ratio.

One way to avoid considering costs in the ranking process is to set up a "conversion table" when first identifying, describing, and ranking risks. This tactic was described in the previous section of this chapter. For example, if the ultimate output metric from your risk-monetization process is likely to be NPV, then a commercial analyst can help you set up a simple conversion table that will translate time or money (1 week—$X NPV, $1,000 Capex = $Y NPV, $1,000 Opex = $Z NPV, and so on) into NPV. In this way, value can be used in the ranking process. An example of this process will be given in Chapter 6.

Regardless of whether you consider cost of impact or value, it is good practice to compare that cost or value measure to the cost of the capture or mitigation action. For example, if a materialized threat will potentially lower the NPV by, say, $2 million but the mitigation action will require the outlay of $4 million over years three to seven of the project, you can use the ever-present project financial spreadsheet—even in deterministic mode—to get an idea of whether or not this $2 million risk should, in fact, be mitigated. If the cost of the mitigation action is "blended" with the impact of the risk, then the ability to decide whether individual risks are "worth it" or not is lost.

Little or No Risk Integration

"Little or No Risk Integration"—the title of this section—would indicate that during the process of risk assessment or management or monetization, integration of risks is a rarity. In the section "Surprise! Integrated Impact Is Almost Always Alarming" of Chapter 1, I describe in some detail and give numerous examples of why integration of risks is critical, and I won't replow all of that ground here. However, given that lack of risk integration is part of the title of this chapter—"Without Risk Monetization, What Typically Is Done"—consideration of risk integration can't here be avoided.

As you will see in later chapters, proper implementation of the risk monetization process necessitates the integration of risks. The dashboard "green light" example and the project start-up-time story in Chapter 1 are but two of a long list of great illustrations of why risk integration is essential to the monetization methodology. If you haven't already, please see the dashboard example in the section "Dashboards and Traffic Lights" of Chapter 1 for a glaring illustration of why integration is necessary.

Risk integration requires several things. First, you have to have something to integrate. That is, you can't integrate risks you have not identified, described, and "quantified" in some way. Therefore, a more or less complete and holistic set of threats and opportunities and the attendant data have to be available for the monetization process.

If you don't monetize, and, therefore, don't integrate a holistic set of risks, then it is exceedingly difficult to otherwise get any realistic estimate of the degree to which risks or combinations of risks add to or detract from the value of the project. Given that increasing value and increasing the chance of success should be the aim of any risk process, it is difficult to argue that monetization is irrelevant. So, if we don't pursue a monetization strategy, integration of a holistic set of risks is unlikely and, as a consequence, huge sums of money and time can be wasted.

Use of the "Boston Square"

One of the most ubiquitous vehicles for expressing risk is the infamous "Boston Square." For numerous reasons, I generally rail against the use of these plots.

Before I enumerate the shortcomings of the Boston Square, I suspect I should first relate to the uninitiated just what is a Boston Square. As I have related many times in the preceding text, most risks—whether they are threats or opportunities—have associated with them at least two properties: probability and consequence. In addition to these attributes, one other thing that most people or companies want to know about a risk is how easy (or cheap) it will be to mitigate or capture. The Boston Square combines these three attributes—probability, impact, and ease of mitigation or capture—in a single plot. Typically, the X-axis of the plot is labeled something like "Ability to Mitigate" because on most Boston Squares are plotted only threats (which is a mistake!).

In Figure 5.1, I show a typical Boston Square. As mentioned above, the X-axis is labeled "Ability to Mitigate or Capture." This will allow the posting of both threats and opportunities on the plot. The X-axis typically (but not necessarily) is divided into three sections—low, medium, and high. Aside from the qualitative—rather than quantitative—X-axis label, this is fine as far as it goes. The real vexing aspect of the Boston Square is the Y-axis.

A Boston Square diagram actually is a three-dimensional plot displayed in two dimensions. The Y-axis metric is the product of probability and impact. Although it is not strictly mathematically "illegal" to generate a multiplication of these two parameters, such multiplication leads to conundrums.

First, but probably least insidious, is the fact that not all threats or opportunities that might be plotted on a Boston Square lend themselves to being part of a multiplication. Qualitative data (high, medium, low, for example) if to be represented on a Boston Square must first be somehow translated into a quantitative metric that can be multiplied by the associated probability (which, by the way, must also be expressed quantitatively). While translation

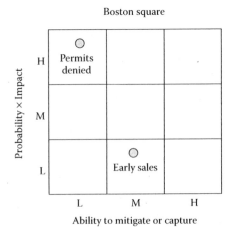

FIGURE 5.1
A typical Boston Square (L = "low," M = "medium," and H = "high").

from a qualitative to a quantitative state is generally desirable (second, however, to collecting the data in a quantitative manner to begin with), the multiplication of the quantitatively expressed risk consequence generates a metric that is at least twice removed from the original expression of risk impact.

A most problematic aspect of the Boston Square is the reduction of two dimensions to one dimension on its Y-axis. There are at least a couple of bad things associated with this dimension reduction. First is the foible, I initially addressed in the section "Threat-Mitigation or Opportunity-Capture Action" in Chapter 2. In that section, I relate a story concerning car-manufacturing companies and airplane manufacturers.

I don't wish to recount the entire story here, but the gist of the tale is that car companies work to reduce the impact of risks. This is evidenced by the vast improvements in, for example, safety features in cars today relative to the cars I drove in the 1960s. Front and side airbags, traction control, antisway technology, vastly improved tires, crumple zones for energy absorption, collapsible steering columns, unibody construction, antilock braking systems, padded dashboards and steering wheels, and so many other improvements are aimed, mainly, at reducing the impact of a collision. That is not to say that the third tail light, halogen headlamps, and other improvements are not focused on accident-probability reduction. However, with crumbling infrastructure ("bad" roads and bridges), more old and young drivers on the road, general highway congestion, and the like, the car and insurance companies know that the likelihood of having a car accident is ever increasing and so most of the effort by auto manufacturers has gone into passing "crash tests" and general impact reduction.

Contrast this focus with that of airplane manufacturers. As I related in Chapter 2, when I first started flying, the ubiquitous Boeing 727 had as safety features some rafts that also served as slides, oxygen masks that dropped from the ceiling, and a seat belt. With minor exceptions like 16-G seats and defibrillators, the safety equipment on the most modern jets is basically the same. This is due to the fact that airlines know that if they fly the aluminum tube into a mountain at nearly 600 miles per hour, no amount of padding, or number of airbags will make much difference. So, the focus of airplane manufacturers, the F.A.A. (in the U.S.) and other agencies has been on lessening the probability of an airplane accident. New spacing requirements, rules regarding flying in possible downdrafts, collision-avoidance radars, and a host of other technologies and regulations all are aimed at preventing the accident from occurring—that is, reducing the probability.

So, what's the point of relating this story? The point is that the X-axis on a Boston Square typically has something to do with "ease of" (sometimes "cost of" etc.) mitigation. I submit to you that it makes a huge difference whether your efforts are aimed at mitigating consequence or mitigating probability. This should be clear from perusal of any plot. However, because probability and impact are combined in the Y-axis, it is not clear to me just what "ease of mitigation" on the X-axis really means. If I were a project manager,

I would demand to know whether our efforts would be directed at reducing the probability of the risk, the consequence of the risk, or both. It matters!

Well, the preceding paragraphs delineate two reasons why I am not a fan of the Boston Squares. However, I have saved the most insidious reason for last.

Y-axis dimension reduction can cause serious risks—threats or opportunities—to "disappear." Because it is less conventional, let's first address an example of an opportunity.

Suppose there existed a potential opportunity to drastically reduce project cost. For a given project, there exists the possibility that a legislative body might pass a law that would completely change compliance issues associated with the project the result being a drastic reduction in project capital expenditure. The downside is twofold. First, the legislation might not pass at all. Secondly, if it does pass, it might not be ratified until after our project is well under way and, therefore, would not have an impact on our project costs.

So, the impact associated with the risk (opportunity) is great but the probability of this opportunity materializing within the project timeframe is very small. When a tiny probability is multiplied by a huge impact, the result is a value that is likely to end up being plotted near the middle of the Y-axis on the Boston Square and, therefore, might draw little attention. If this opportunity could be a real game changer for the project and the fortunes of the company, such a risk should plot in a place on the Boston Square in which it would draw attention.

Threats are no different. If there is a threat that could potentially be disastrous for the project and the company (high impact) but such a threat is not highly likely to happen (low probability), due to the multiplication of probability and impact such a threat will also end up plotting near other legitimately unalarming risks near the middle of the Y-axis. I know of no deciding body that would not desire that such potentially disastrous events not be paid special attention and highlighted on any display. The Boston Square does not facilitate this.

Well, I'm a great believer in the philosophy that one should not just complain about something, but that one should offer a potential remedy to address the complaint. As a counter to the Boston Square, in Chapter 6, I will propose and make examples of using Probability/Impact Grids, Probability/Mitigation/Capture plots, and Impact/Mitigation/Capture diagrams—but not here.

Data Gathered from Those with Vested Interest

This subject is always a sticky one to address. It is such because it is difficult to illuminate the problem without putting into question the basic honesty of individuals and groups. To be honest myself, risk monetization does not represent the total remedy for this issue, but the cold-calculation, quantification, and documentation aspects of the monetization process certainly tend to alleviate the condition.

Fundamental to the issue is the internal organization of most large companies and their approach to staffing project teams. Especially, but not exclusively, when a project requires the relocation of company personnel, the "vested interest" monster raises its ugly head.

When relocation of company personnel and their families is associated with successful execution of a project, it is typical that the people involved will acquire housing, put their kids in schools, establish a network of friends, and do all of the things attendant to a "move." It is common for these people to feel as though their success in the company—and, perhaps, the existence of their jobs—to the success of the project.

Years ago, when the world was young, I had the rather unpleasant task of traveling to a foreign country to perform a financial risk analysis on a major project. The purpose of the analysis was to determine, in a quantitative and objective manner, whether or not the already sanctioned project should be allowed to continue.

Of course, many of the individuals from whom I was required to glean information and data for the assessment were expatriates who had, as described above, obtained housing, established social networks, enrolled their kids in schools, and so on. Gathering information from such individuals is a bit like asking the grandmother whether or not the baby is cute. Just what did you think she was going to say? Like the grandmother, those with a vested interest in the project are not likely sources of the most objective views. Another approach had to be taken to data collection.

Given that I was tasked with generating an objective and fact-based assessment of project financial viability, I quickly determined that querying the expatriate population was not the road to success. This was the first time in my career that I thought about monetization in the sense of needing to tie the assessment input data back to "ground truth." The solution was to minimize the interview process for collecting data and to travel to actual production sites, logistics hubs, and other primary locations. At those locations, I could obtain "raw" data that could then be utilized in the risk assessment. It certainly takes a lot longer to do this and such practice does not endear you to the project personnel, but it is a necessary step.

So, although risk monetization is not the be-all-and-end-all for fixing the "vested interest" problem, the aspect of monetization that ties data back to "ground truth" and that utilizes quantitative and objective information is at least a hedge against biased information. Without a comprehensive and consistent monetization practice, information from those with a vested interest can typically be that which is reported and upon which decisions are made.

Faulty Forecasts

Although seemingly similar to the "vested interest" issue described above, the malady of faulty forecasts can be vexing even when vested interest is not a factor. Sans vested interest, such errant prognostications can result from, just for example, a lack of skill when identifying risks.

You will note that in Chapter 3 of this book, I include both threats and opportunities in my definition of the term "risk." I'm sure that many of you reading this book might still not agree that opportunities should be considered as risks. However, not considering them can lead to trouble.

Those who reject the idea that opportunities are a type of risk, or, those who have not been exposed to that concept will focus risk-identification exercises on threats. It is not that these folks are unaware of opportunities, but when upside is not formally addressed in the risk-identification process and given equal footing with threats, then there is a great propensity for a risk analysis to be "top heavy" with the consequences of threats. This is especially true when the risks—mainly threats—are translated into impact on value.

As part of a cogent risk-monetization process, opportunities are given equal emphasis. Translation of the opportunities into impact on value—along with the threats—always yields a risk-based forecast that better fits with reality. Always be sure to include the identification and definition of opportunities as part of your risk identification and, in turn, your risk-monetization process. Examples of this will be given in Chapters 6 and 7.

A second common path to faulty forecasts does, in fact, have to do with vested interest. This is the antithesis of the problem described in the immediately preceding paragraphs.

When a project is envisioned, it is not uncommon that, even before the RIW is held, a commercial analyst is asked to produce a financial model for the project. This model usually takes the form of a spreadsheet. The model is faithfully created, populated with data, and run. Results from the model are commonly accepted as the "base case" for the project's financial forecast.

Next, a RIW might be held. It is common practice, as I have described earlier in this book, to have either the project manager or the commercial analyst open such workshops by describing just what constitutes the project. This is done so that the legal representative, the negotiator, the construction engineer, and others are not all envisioning different projects—even though they are all in attendance to serve the same project. The project described at the opening of the RIW is the "base case" and is the foundation of the commercial analyst's financial model.

One objective of the risk-identification process is to determine and define risks—both threats and opportunities—that are not already considered in the base case. Unless the RIW is convened prior to the generation of the base-case model, it is very common that a long list of threats is brought to light and very few opportunities are identified.

This phenomenon is ubiquitous because it is typical that nearly all opportunities—even those with relatively low associated probabilities—have already been accounted for in the base case. Project teams can be in competition with other teams for sanction of their project. Therefore, it behooves those teams to construct a base case that is as "rosy" as is practical. Forecasts from such base-case models are, as you might guess, a bit overly optimistic. It is only through the rigorous risk-monetization process that such lofty projections are brought down to earth. Examples of this will be given later in this book.

Lack of Organization

Although this section does not follow the "Incomplete Representation" rendition in this chapter, there certainly are conceptual links between the sections. A main connection relates to the organizational test of wills.

In Chapter 2, I describe in detail some of the salient attributes of the typical risk register. If there exists for a project a single risk register that contains the threats and opportunities emanating from every discipline, it is a difficult enough task to establish who will control and update the register and who will enforce the timely execution of threat-mitigation and opportunity-capture tasks. This job is many times more arduous when disparate disciplines each insist on keeping colloquial registers.

When I say the words "risk-monetization process," people who hear those words tend to assume that such a technique is mainly about translating risks into numbers, dollars, and the like. While such translation certainly is part of the risk-monetization process, it is just a part.

As will be delineated in detail in Chapter 6, much of the risk-monetization process involves creating an organization in which monetization tactics can exist. As you might guess, there exist many facets to this, and some of them can become quite contentious.

One Lump or Two (or More)?

This is one of those situations in which I am cursed with an embarrassment of riches by way of examples. Psychologists will tell you that a major source of anxiety for most people is a true or perceived lack of control. So it is with organizations. It is exceedingly characteristic that every major discipline associated with a project—legal, commercial, engineering, finance, health and safety, etc.—will argue that they should each create and maintain their own register of risks. Confidentiality, uniqueness, discipline-specific knowledge ("Nobody else would know how to handle our stuff ...") and other arguments are put forward in defense of keeping a separate register.

However, if the risk-management process is to be successful for the overall project, then all of the risks should be kept in a single register, or, in separate registers that are real time and dynamically linked so that they appear to be a single register. This is one of those times—among many—when the anointed RPP has to have the backing of project management to compel individual-discipline leaders to comply with the single-register (or, single-register image) notion. For efficient project execution and conservation of expenditure, all risks need to be quantified and ranked (see the section "One, Two, Three—GO!"). This is among the first tests of wills that will accompany the implementation of any risk-monetization process.

Who Has Access?

Imagine that the RPP prevails and all risks exist in a single register. Now the anxiety level of denizens of the various discipline departments really goes off the chart because access to the risk register has to be restricted—usually to just the RPP.

Imagine being the RPP who is charged with ensuring project management that all of the attendant risks have been identified and that responsible parties will faithfully and in a timely manner execute the threat-mitigation or opportunity-capture plans delineated for each risk. Further imagine that anyone—including those who have responsibility for executing the mitigation or capture tasks—can access and modify the register entries. Such ad hoc modification of the register data makes it impossible for the RPP to know if actions have actually been completed. Such ignorance makes it, in turn, impossible for the RPP to assure management that risks are being addressed.

So, the only practical way for the RPP to "cover his or her posterior" in a properly implemented risk-monetization process is for the RPP to act as the sole individual who can modify—in any way—the risk register information. This, of course, leads to yet another raft of contentious issues.

Who's the Boss?

If the RPP has sole custody of the risk register, then when a party responsible for executing a threat-mitigation or opportunity-capture plan comes to the RPP and says: "Yeah, I did that yesterday—change the risk register to reflect the fact that I met my obligation," does the RPP take every person at their word? I suspect such an honor system would not work in the best interest of the project.

Of course, the RPP would need to see some physical evidence or documentation that would substantiate the claim of task completion. Not only do people find it distasteful to have their statements of compliance questioned, but it really rubs them wrong when someone outside of their discipline management can demand that they document the achievement. If such documentation is not forthcoming, then the RPP has no choice but to alert the project manager that the risk is yet outstanding. This makes the RPP a "real popular guy," but it can't be helped.

How People Are Alerted

Suppose you are the project RPP. You have responsibility to see to it that, literally, hundreds of risks are appropriately addressed by dozens of individuals—none of whom formally report to you. Along with these hundreds of risks are hundreds of "due dates" related to the time by which particular risk-action plans should be completed. It is practically impossible for any individual to keep abreast of all of the dates and actions. Even if it were possible to somehow dynamically "know" all of these things, it is even more impractical to imagine that the RPP is going to physically contact each responsible party—some of whom might not be on the same continent.

This scenario makes the argument that any attempt to keep a major project risk register in a spreadsheet on a particular hard drive is folly. Any cogent risk-monetization plan will call for an online register that automatically can send e-mails or other types of electronic alerts to responsible parties. The timing of the alerts can be set by the RPP, but they decidedly should not only be just before the due date. Sending too many alerts will cause them to be ignored by the receiving party. Alerts that are too late are practically useless. A computer program that can send out well-timed alerts to the responsible parties—with a copy of the alert to the RPP—is a risk-monetization-process practical methodology.

The Importance of Keeping It Fresh

Thus far, all discussion of risks in the risk register has assumed that those risks emanated from a facilitated risk-identification event held early in the project's life. I have also painted a picture of a risk register full of threats and opportunities and that, over time, mitigation or capture actions would be successfully completed and, therefore, the risk would be marked as "inactive" in the register. That's all very nice, but given that most projects can take years to complete, is it really a rational assumption that new risks—both threats and opportunities—will not crop up throughout the lifecycle of the project? Of course, they will.

Given that new risks will most assuredly arise, just how does the RPP capture those new risks in the register? This can be a tricky thing.

Until the team members representing the various disciplines get comfortable with the idea that they are, in some ways, subservient to the RPP, the RPP is not going to be the most popular person in town, so to speak. Discipline experts might have capitulated in attending an initial project RIW during which the initial risk register was populated, but such individuals are likely to be reticent to attend one or more such workshops throughout the lifetime of the project. Besides, the RPP needs to capture new risks as they come up—not at episodic events. In the section "Gatekeeper or Free-for-All?" of Chapter 2, I outlined a few simple but effective methods for "keeping it fresh."

Mitigation of Threats

In the section "Mitigation and Capture" of Chapter 3, I present my definition of the term "mitigation." Also, in the section "Faulty Forecasts," I describe the base case that contains most or all of the opportunities—even the ones with associated low probabilities—so that a subsequent RIW produces a long litany of threats but very few new opportunities that have not already been accounted for in the base case. Given that I have addressed the threat-mitigation and opportunity-capture issue at least twice before, you'd think that I'd have exhausted my comments on the subject—but you'd be wrong.

There's at least one more thing I will address regarding threat mitigation. That is, it's the popular thing to do. My point here is that it is commonplace for project teams to present to project management a long list of potential problems (threats) and the actions that will, hopefully, mitigate those issues. Some of the propensity for this behavior stems from the fact that project management seems to expect that lists of risks—in the form of threats—is what they want to see. The focus-on-threats mentality also emanates from the desire on the part of, say, project engineers, to be perceived as "problem solvers." Most of the formal education offered to such individuals focuses on the generation of solutions to problems.

While it is right and noble to present solutions to problems, it is not seen as "sexy" the presentation of opportunities and plans to capture them. It just isn't. This is a major mistake in the typical risk assessment or management process. Identification and capture of major opportunities is not only beneficial to the cost and schedule of the project, but the capture of a major opportunity might actually be, directly or inadvertently, a multi-pronged mitigation action that will offset a host of threats. This calls for an example.

Suppose that a project team was considering a particular right-of-way for a rail line. The considered route was selected because it was already being used as a pipeline route and many of the right-of-way issues had already been addressed. The considered route, however, has some nasty geographic attributes and a burgeoning local population is encroaching on the route. Many mitigation plans have been established to offset threats related to potential problems with the pipeline and with rail interactions with population centers.

However, a major land owner in the area has recently died. His heirs are open to discussing right-of-way for the rail line. Because the land was in private hands, it does not contain population centers or pipelines with which the rail system would have to contend. The passing of the land owner is an event that is post project start up, but if that land could be procured, many expensive-to-mitigate threats along the existing route would be moot.

Reading the preceding paragraph, you might think: "Well, sure, it only makes sense that the team should look into procuring rights to the land ..." and you'd be right. However, I can tell you from years of personal experience that once a project has started down a particular path—in this case, the rail following the pipeline pathway—it can be exceedingly difficult for a project team to change a major part of the project even though the change would bring enormous good to the effort. In addition, it is not the nature of most of those working on the project to take their eyes off of the tasks at hand to investigate potential opportunities that were not originally associated with the initial project plan. Also, if you are, for example, an engineer on the project, what business is it of yours to be looking around for alternate routes and to be taking time to talk to kin of the land owner? These types of things—glaringly obvious opportunities—most times are not identified or captured once the project is "in full swing." A well-designed risk-monetization process, as part of a risk-management process, will always make opportunity identification and capture an "equal partner" with threat mitigation.

Still another aspect of threat mitigation is the mindset that "mitigation" means what we should do about a threat after it materializes. As I stated previously, many of the manuals in a Safety Engineer's office are full of advice regarding what to do if particular incidents transpire. As I indicated in Chapter 3, the term "mitigation" in the risk-monetization context is going to translate to what we do to prevent the threat from materializing in the first place. This is a significant change in mindset for most project personnel.

Initial Budget Estimates

In the section entitled "We'd Like to Keep Two Sets of Books—The EVS and EVP," of Chapter 1, I describe briefly just what these two terms mean and why they are important. In a nutshell, the EVS (expected value of success) is the value about which a project team plans. For example, if a factory is likely to produce one million tons of product in the coming year, then logistics personnel have to plan to rent enough train cars to move one million tons of product. However, the plant will only produce one million tons if the upcoming negotiations with the union do not break down and a strike is called. There might exist, say, a 50% chance of such a production stoppage for up to half the year.

Well, the logistics folks have to sign contracts for train cars early in the year. Therefore, they have to commit to the 1 million tons capacity with a stiff penalty if they renege on the contract. However, when production from this plant is compared with other plants, or, when a roll-up of corporate-wide production from many plants is performed, then the value that would be used is the EVP or the expected value for the portfolio. In this grossly over-simplified case, the EVP would be 500,000 tons (0.5 × 1 million).

It is not uncommon at all for companies to establish the EVS values for projects—projects that have less than a 100% chance of attaining the EVS value—and to "add 'em up'" for corporate forecasts. Equally faulty is the practice to rank, for example, projects based on the EVS without taking into consideration the probability that any given project will actually attain that value.

Initial Schedule Estimates

In the section "It's about Time" of Chapter 1, I address the importance of integrating schedule risks with other risks in a project. The focus of this section will be decidedly different.

As the old adage goes, time is money. My focus in this book tends to be on the translation of threats and opportunities into dollars. Even though in a real-world project setting you would see many Gant charts (plots of schedule), you won't see any here. That's because the Gant chart is just a "half way house" between risks (events) that would impact the timing of the project and the financial bottom line. Again, time is money.

Having said that, there is at least one schedule-related problem that warrants mention here. The RPP needs to be an individual who possesses the skills necessary to integrate both physical and time-related risks. If the RPP is not versed in the language and methods of time analysis and how it relates to the physical aspects of the project, then that RPP should seek the assistance of someone with such abilities.

It never fails that at the very beginning of a project, a Gant chart is built. On such a chart, horizontal bars represent the projected length of time taken by each project step. A stylized Gant chart is shown in Figure 5.2.

It is part of the remit of the RPP to ensure that the timing of project steps and the threat-mitigation and opportunity-capture actions in the risk register are in sync and are correlated. For example, one of the risks in the register might relate to an opportunity to purchase compressors in the current fall season to be used the following summer. Those compressors, if available before this winter, will have to be stored at the project site for early use in the summer. Early purchase and storage of compressors would significantly advance the project schedule relative to waiting until spring or summer to purchase compressors and have them shipped to the project site.

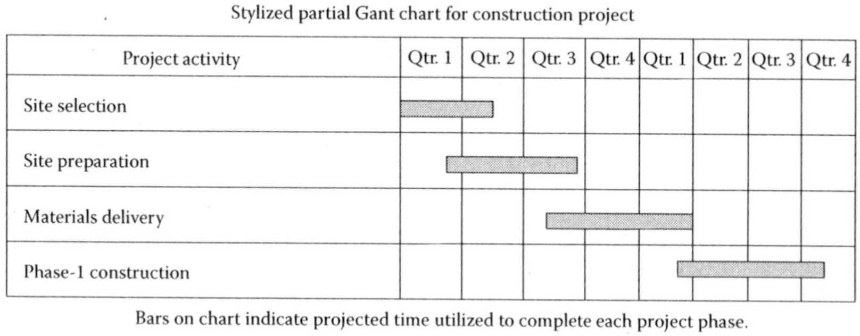
Stylized partial Gant chart for construction project
Project activity
Qtr. 1
Qtr. 2
Qtr. 3
Qtr. 4
Qtr. 1
Qtr. 2
Qtr. 3
Qtr. 4
Site selection
Site preparation
Materials delivery
Phase-1 construction
Bars on chart indicate projected time utilized to complete each project phase.

FIGURE 5.2
Depiction of a partial and stylized Gant chart for a hypothetical construction project.

A Gant chart might show the use of the compressors in early summer. However, if the opportunity to early-purchase the compressors and the ability to store them at the project site over the winter does not materialize, then the timing of the entire project on the Gant chart will slide forward in time. It is part of the RPP's job to account for the possibility that the equipment might not be available before next summer and, therefore, the RPP must make decision makers and project management aware that the timing represented in the current Gant chart could be overly optimistic. In fact, probability-weighted adjustments to the Gant chart—reflecting the possibility of failure to capture the compressor opportunity—should be made immediately.

Such adjustments are a difficult sell. Project team members are anxious to meet their time and budget constraints. Project management is even more apprehensive. The RPP, seeing the potential incompatibility between the Gant-chart timing and the opportunity-capture probability in the risk register, might be the only advocate for official adjustment of the schedule so that it better reflects reality. This is a real problem. If the RPP does not have the backing of company management to make adjustments to project projections—or at least to advocate such adjustments—then much time and money can be wasted by project teams working on "phantom" projects.

Another real problem related to schedules is the propensity for project management to override the meticulously considered timeline created by the project team. I can't relate to you the number of times I have witnessed project teams present to project management what the team considered to be an accurate projection of how long the project will take. I kid you not when I tell you that I have observed a project manager walk over to the Gant chart and put a pin in the timeline and indicate: "I understand your reasoning for the timeline you present, but the project will be completed by this date ..." indicating where he or she had inserted the pin part way along the project team's timeline. Sound familiar?

It is not that I think that the risk-monetization process can alleviate this malady. Nor can the RPP believe that she or he can throw themselves in front of the train, so to speak, to successfully counteract such practices. However, if proper quantification of both impact (time) and probability have been documented, it is the role of the RPP to, along with project-team-member support, to challenge the seemingly arbitrary contraction of the project time line in light of the integration of real-world projections. Again, backing of the RPP by company management is essential.

Sensitivity

Although many of the critical facets of risk assessment or management or monetization have to do with psychological aspects, this is not one of them. In this book, the term "sensitivity" will relate to the relative impact that individual input parameters have on the major output parameters. That is, a sensitivity analysis helps someone determine which input parameters most influence the output.

To those not versed in the art of sensitivity analysis, the practice of determining which inputs are most influential might be deemed academic or, worse yet, easy. Trust me, it is neither. As a boneheaded example, consider two input parameters—A and B. Parameter A is a variable with a range (represented, say, as a distribution) from 2 to 6. Variable B has a range from 1,000 to 1,000,000.

As part of this example, let's consider an output parameter C, which is the sum of A and B. That is, $C = A + B$. Which input (A or B) parameter has the greatest influence on the magnitude of the C coefficient? Clearly, it is B. If we removed A from the equation, the impact on C practically not be noticeable.

Now, let's consider the same three parameters—A, B, and C—but a different equation relating them. Our equation is now, $C = B^A$ (C = B to the Ath power). Now which input parameter has the biggest impact on C? I submit that B squared is much different than B to the 6th power.

The point of this simple example is to emphasize that you cannot necessarily deduce—from the magnitude of the input-parameter coefficients—which input parameters will most influence the calculated output. Except in the situation of a simple sum, you must know how those input parameters are combined to make such a deduction.

If you can't generally determine the sensitivity of output parameters to input variables—and you can't—then how can you figure out which of the input parameters are most influential? Perhaps more importantly, why would you want to know this? To address the second question first—why you want to know this—I will here relate to you one of my many real-life war stories.

Once upon a time, I was called to another city by a commercial analyst who was charged with the task of determining whether or not the company for which they worked should purchase the product-handling capacity of a relatively small competitor. To analyze the problem, the analyst had constructed an exceedingly complex set of spreadsheets. First requested of me was to aid in the establishing of ranges for various input parameters so that the spreadsheet could be used as the basis for a Monte Carlo analysis. I did this.

Sometime later, I was recalled to the analyst's office so that he could show me how the calculated result—in this case, NPV changed with various levels of purchased (increased) volume of product. The analyst had meticulously plotted all of the increased-volume amounts against financial impact and was about to take the results to management.

Remember, this is a true story. While I was sitting there looking at the plotted results, the commercial analyst was called away. Me being me, I couldn't resist tampering with the model and wrote some code that would generate a Spearman Rank Correlation (SRC) (to be discussed later in this book) analysis. When the analyst returned, I disclosed that the analysis indicated that, relatively speaking, increases in purchased volume had little impact on the resulting NPV. This was mainly because the cost of purchase nearly negated the increase of revenue generated by the increased volumes.

A sensitivity plot showed that other parameters such as timing, product type, logistical aspects, and other parameters had significantly bigger impacts on NPV than did increased volume. The analyst was surprised at this and called his boss over to discuss it. I then departed and to this day, I don't know what they decided to do, but I hope it was not to purchase more volume of just any type!

I hope the diatribe above makes clear why you would want to know which variables are most influential. The commercial analyst, his boss, and others assumed that the purchase of additional volume would most impact the NPV of the project. As it turned out, the impact of the additional volume on the financial viability of the purchase of the small competitor paled in comparison to the influence of other parameters. Had I not generated the sensitivity analysis, the commercial analyst would have based the purchase or no-purchase decision on erroneous assumptions.

Well, so much for why you might want to know which parameters most influence an output parameter. Now to the question of how this is done. Turns out, there is an evil way and a good way to do this.

First, let's look at the evil way. Way back when Monte Carlo analysis was first being applied to physics problems (the Manhattan Project during World War II), it became clear that it might be useful to know which input parameters were having the "biggest" impact on the calculated results. The basic problem is that if all input variables are offering up a new value on each iteration of the Monte Carlo process, how do you get a grip on which one is having the most significant impact on the calculated result (see my example of the variables A, B, and C in the discussion above)?

Some quasiclever individual—and I have no idea who that was—decided that the solution was straightforward. If you have, for example, five input variables all represented as distributions, then the way to determine which variable is having the "biggest" impact on the calculated output is to hold all variables—except one—at a constant value—the mean of the parameter range, for example. Then, when the Monte Carlo analysis is run, the range generated for the output parameter could only come from the one parameter that was not held constant.

If the range in the output parameter is recorded and represented as a bar on a plot, then that is the influence of the one "ranging" input variable on the output parameter. If this process is repeated for each input variable—that is, in turn, each input parameter is allowed to "range" while all others are held constant and the results from those individual experiments are represented as sorted bars on a common plot—then you get the classical "tornado" diagram. Such a tornado diagram is shown in Figure 5.3.

Well, it should be evident to even the most remedial or readers that there are bodacious holes in the logic that produces plots such as that shown in Figure 5.3. Just one hole in the logic that produces such plots—and there are many more holes that I won't address here—is the fact that many parameters in an analysis are correlated.

Suppose you were considering a relatively complex set of equations that, among many others, used the parameters Depth and Temperature as inputs. These two variables are each represented by distributions. That is, they are not deterministic in nature and are represented by a range of values. Depth,

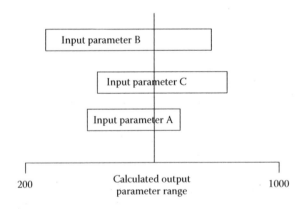

The width of the bar for each input parameter represents the change in the value of the output parameter when that input parameter is allowed to take on the values representing its range and all other parameters are held at a constant value. Except in special cases, this is a highly flawed approach to determining the impact of input parameters on the calculated output parameter.

FIGURE 5.3
Example of the ubiquitous but not recommended "tornado plot" created by holding all parameters constant—save one—in a Monte Carlo analysis (see text for details). Such a plot is intended to gauge the impact of individual input parameters on a calculated output parameter.

for example, can range from 0 feet below the surface of the earth to 20,000 feet below the surface. Temperature can range from 60 to 500°F.

Depth and temperature are correlated. That is, the deeper in the earth you go, the hotter it gets. So, a plot of Depth against Temperature would look something like that shown in Figure 5.4.

When both Depth and Temperature are allowed to vary in the Monte Carlo process, a correlation algorithm ensures that reasonable pairs of values are selected. For example, if a relatively great depth is randomly chosen from the Depth range, then the correlation algorithm selects a relatively high value from the Temperature range. Conversely, if a relatively shallow depth is selected, then a relatively and concomitantly low temperature is selected. Correlation of the values ensures that "goofy" combinations of, in this simple case, Depth and Temperature are not combined in the equations.

If we apply the "hold all parameters constant except one" philosophy, then considering the correlated Depth and Temperature input variables, we would generate mainly bogus results. Let's say that you first wanted to test the influence of Depth on the calculated output. In that case, you would hold Temperature constant at some value. It does not matter what Temperature value you select as the constant—any one value will yield equally silly results. For example, if you select the mean of the Temperature parameter range, then all Depth values will be combined with that mean Fahrenheit value. Given that Depth and Temperature are correlated in real life, almost every iteration of the Monte Carlo process would utilize a bogus combination of Depth and

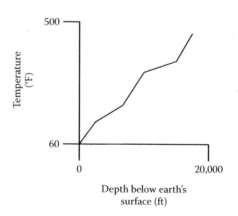

Stylized relationship between depth and temperature. As depth increases, temperature increases. These parameters, therefore, are correlated.

FIGURE 5.4

Example of two correlated parameters (Depth and Temperature). Such a relationship between two or more parameters invalidates the practice of allowing one parameter to take on a range of values while holding constant the values of other input parameters with which the "ranging" parameter has a relationship.

Temperature. Could you then plot the range of the output parameter on the tornado plot and believe that the bar represents the impact of Depth on the calculated output? Of course not.

To remedy this foible, a process called SRC is applied. It is not within the scope of this book to delve into the internal workings of SRC, but suffice it to say that SRC is a nonparametric (distribution free) method that facilitates application of correlation while allowing all variables to vary simultaneously. Correspondence between individual input parameters and calculated outputs are generated by comparison of ranks. The reader can easily look up SRC on any internet search engine to read about the details of how it works. An abbreviated SRC plot for a construction project is shown in Figure 5.5.

In Figure 5.5, the X-axis is correlation (range from −1 to 0 to +1). The length of the bars represents the relative degree to which the individual input parameters influence the output parameter. Bars extending to the right of the center line are positively correlated with the output parameter, while bars extending to the left of center have a negative correlation with the output variable. In Figure 5.5, the % value associated with each input parameter indicates that parameter's contribution to the calculated (output) parameter's range (contribution to variance, in actuality).

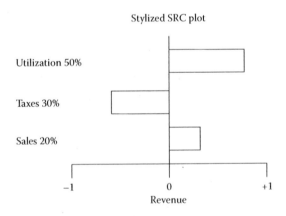

A stylized SRC plot. The % value associated with an input parameter indicates the contribution to the calculated output parameter's (revenue, in this case) range of values. The length of the bar associated with an input parameter indicates the degree to which that input parameter correlates with the output parameter. Bars extending to the right of the center line indicate positive correlation, while bars extending to the left indicate negative correlation.

FIGURE 5.5
A stylized SRC plot. The SRC approach facilitates determination of the contribution of each input parameter to the variance of the calculated output parameter while allowing all input parameters to take on a range of values. See text for details.

Risk Interpreted as Uncertainty

In the section "A Focus on Success and Value—*Not* on Reducing Uncertainty" of Chapter 1 and in the section "Not Primarily about Minimizing Uncertainty" of Chapter 3, I relate examples of individuals who are responsible for project success. In both examples, the point was that efforts to increase the probability of success for a project might result in an increase in uncertainty—and that's OK.

Emphasis in this section is different. Here, I will relate, as briefly as is practical, the tendency of project teams to interpret "risk" as "uncertainty." That is, some project teams deem that when they have put ranges around otherwise deterministic values and have run Monte Carlo analyses on those ranges, the risk work is done.

I know I say this a lot, but I can't tell you the number of times that I have interacted with project teams the members of which were of the opinion that risk analysis constitutes putting ranges around deterministic values—those values usually existing in a spreadsheet. Popular spreadsheet add-on programs lend false credence to this philosophy.

Risks are events. Most risk analyses should be focused on increasing the chance of success (or, conversely, decreasing the chance of failure) for an endeavor. In fact, when I teach classes one of my common refrains is that if I were in competition with another individual regarding the execution of a risk analysis and if I had, as a handicap, to give up one aspect of the risk process, I would certainly sacrifice the ability to analyze uncertainty. Even with this handicap, I would likely "beat the pants" off of the competitor by adroitly implementing the risk identification, risk ranking, mitigation or capture actions, and other steps associated with a cogent risk analysis. Almost always (but not always) the analysis of uncertainty is secondary.

This philosophy is a tough sell to the "quants" who typically get assigned to perform "risk assessments" for projects. Their usual approach is simply to use a spreadsheet add-on program to put ranges around input parameters, correlate a few parameters, and then crank out all sorts of cool-looking plots that "prove" that the risk analysis has been done. In my humble opinion, none of the risk analysis has been done when only such steps are taken. Examples of how to actually carry out a risk analysis will be the subject of subsequent chapters.

Poor Risk Communication

In Chapter 1, I present a section entitled "Communication." In that subsection, I mainly, but not exclusively, focus on the necessity to come to consensus—across the project landscape—on the definition of terms. In this section, I will present yet another vexing attribute of communicating—that of the chain of communication.

Many times, perception is reality. That is, the perception of the successful execution of a project becomes the reality even though that perception does not represent the real-world success profile of the project. Unwarranted exuberance regarding perceived project success can lead to completely erroneous and sometimes destructive decisions.

In an attempt to keep this from happening, it is critical that the RPP establish—from the very beginning of the project—just who will communicate what to whom. Untrue and conflicting reports regarding the risk aspects of the project can cause organizations to not only operate under false pretense, but can cause real damage to project success.

Just for example, there might exist two organizations—logistics and construction—that are partially responsible for preparing a remote site for construction of a facility. Regarding the readiness of access routes to the construction site, if overly optimistic reports of threat mitigation or opportunity capture from the logistics team are erroneously reported, it might cause the construction department to prematurely purchase and begin to move material. This, of course, can cause havoc with budgets and schedules.

It is, therefore, critical that communication between project subteams and communication to higher level decision makers be managed by the RPP. This, of course, can result in tension in the workplace. Just another example of why, in a well-executed risk-monetization or risk-management effort, the RPP must enjoy real backing by upper-level management.

Piecemeal Presentations

This section might seem to have overlap with the section "Poor Risk Communication" immediately preceding, but the following dialog really does address a separate problem.

Though not exclusive to situations in which a RPP has not been assigned (or, one has been assigned but is anemic in his or her power), the piecemeal presentation process is more prevalent when there is no hand at the helm. The problems described in this section can vex project success regardless of whether or not the RPP has been successful in establishing lines of communication.

Piecemeal presentations are mainly, but not exclusively, a temporal issue. Here's what I mean by "piecemeal presentations."

Let's suppose that there is no authoritative ramrod assigned to the risk aspects of the project, or, that an RPP has been appointed but that the position is a ceremonial one and that the RPP is, essentially, a toothless police dog. Let's further surmise that there is one office—usually someone high up the chain and very distracted, like a Chief Financial Officer (CFO)—to which risk reports are to be delivered. Of course, those reports would not be put directly on the CFO's desk but would likely initially land in the in-box of an administrative assistant.

Let's imagine too that there are six main departments from which risk reports are due: Legal, Commercial, Finance, Security, Logistics, and Construction (engineering). Even if a firm due date is set for risk-report delivery, there are departments that will—reflecting mainly the personal attributes of their leaders—view such a report as something to get out of the way as soon as possible. Those departments will deliver the reports early. Of course, some leaders are procrastinators and the attending reports will be delivered from them on the due date if not late.

Any set of expressions of risk are "snap shots" of the current situation. That is, they always reflect the realities of the world at the time of their construction. When time is allowed to pass between the issuance of reports, it is folly to expect that such reports universally represent the world situation on the due date. This reality leads to extreme difficulty in integrating the risk reviews because some are outdated and some have been done in such haste (due-date eve.) that it is meaningless to attempt coalescence—and it is an integrated picture is what is required.

Requiring a particular format for the reports typically fails to help. Turns out, it is not the format of the presentations that prevents meaningful integration of the information, but rather it is a fundamental lack of common understanding of the risk-related terms and definitions that precludes essential confluence of the data.

I regard the scenario described above as typical of what happens when a well-conceived risk-monetization or risk-management plan is not enacted. I have already partly described the method of a "brainstorming workshop" that can be an essential part of the monetization process. Later in the book, I will describe other processes that can serve a similar purpose. So, regardless of the data-collection method used, one of the common traits regarding those tactics is that they all attempt to simultaneously represent the current state of the situation and that current state—the "snap shot"—is gleaned from all pertinent disciplines in a temporally satisfactory manner. Also, such techniques promote the use of common language and understanding of the type and format of the information to be collected.

Determining the Reduction in Probability and/or Impact

Given that I will in ensuing chapters address how modification of initial estimates of probability and consequence can be accomplished when monetization is part of the process, I will here disclose at least one commonly used method to change initial estimates when quantification and monetization have not been employed.

Now, I can hear some of you screaming that quantification and monetization are hardly the same thing and that in many risk exercises, probabilities and impacts are quantified but not truly monetized. No argument. However, it has been my experience that truly meaningful initial estimates of probability and/or impact and subsequent elevation or diminution of those estimates is most practically and efficiently done when monetization is in the mix.

That is not the case at hand. The enigma here is, of course, "What is typically done when risk monetization is not in the mix?" That is, just how do we make initial-estimate modifications sans monetization? This is best illustrated by example.

First, I would ask you to note that this has little or nothing to do with whether we have represented our probability or consequence in deterministic (single-valued) format or in an uncertain manner (as a distribution, for example). Next, I would request that you follow me through this elementary scenario.

My company is attempting to build a manufacturing facility in a foreign land. To win the contract, the company had to agree to take control of an already existing factory—one built by my company's competitor and with which the host government is exceedingly dissatisfied.

Government annoyance stems from the fact that the pits that necessarily and temporarily hold process waste, in the form of a liquid, were not lined with nonpenetrable industry-standard materials (clay and/or a plastic liner). Leakage of waste from the pits has contaminated the drinking water in nearby villages. The government seized the infrastructure, paid the competitor company to leave, and suspended operation of the plant until a suitable and more responsible company with domain expertise could be identified and engaged. That would be my company.

So, under the agreement, my company will operate both the older competitor-built plant and our newly built factory until such time that we can phase out the older facility. In the new-plant plans submitted to the government, my company, as always, committed to the construction of waste pits that were lined both with impenetrable (to liquid waste) clay and double-the-required-thickness plastic liners.

A newly appointed Minister of Environmental Affairs (MEA) for the country has reviewed the submitted plans and is holding up issuance of critical permits—for perhaps as long as a year—because he is requiring that the waste pits be located miles from the manufacturing facility and from villages. It is the opinion of our company that he either believes that we built the original offending pits (because we now operate them), or, that he assumes that we will act just as did our competitors when it comes to pit construction. Neither of these things is true.

A rerun of our risk model showed that the costs of waiting the extra year and the remote location of the pits would ruin the economics of the project and would, essentially, kill it. We have estimated that if this threat is not mitigated, the probability that the threat will kill the project is 90%.

I'm the company's environmental expert and I was charged with creating the succinct materials that I was to present directly to the MEA so he might realize that neither delay nor remote location of the pits was necessary. Because the MEA is new in the position, does not have much background in such matters, and is under some pressure from his own government to put people back to work, my company management believed that a friendly, expert, and direct presentation of the facts would mitigate the situation. That is, corporate leaders postulated that such candid interaction with the MEA would reduce to zero the probability of a delay or a remote location for the pits.

I did put together a succinct and convincing argument in a very brief document. I also arranged a meeting at the office of the MEA. However, upon arrival, I was shown to a conference room in which I met—for only 10 minutes—with an underling of the MEA. It was clear to me that this person had little or no training in environmental science. He requested only a brief description of what was in the document and, without ever opening the binder, thanked me for my efforts and assured me that he would personally pass this along to the MEA and that we would receive a reply from the MEA "at an appropriate time"—whatever that meant!

When I returned to the office, budgeting and planning for the project were in full swing. When I related my experience to project managers, they realized that the probability of abject failure of project on environmental grounds was likely no longer 90%, but it certainly was not zero. Chance-of-failure estimates and probability-impacted economic projections for the project were due to be delivered to corporate headquarters in less than a week. Management looked to me for a solution to the dilemma.

Well, here's at least one place where it gets tricky. Until now, we had all considered that the consequence would be a 1-year delay. What if that is not so? Commercial planners and financial analysts assured me that a 1-year hiatus certainly would torpedo the project. "What about a 6-month delay?" I thought, or any less-than-1-year timeframe?

To get a grip on this, I was directed to arrange a meeting with our construction team. In a conversation with them, it was made clear to me that construction equipment could only be practically transported to the proposed building site during the relatively dry—with respect to rainfall—late summer months. Those months were coming up soon. So, if we suffered more than a 2-month delay, that was essentially the same as a 1-year delay (waiting until the next relatively dry season). This indicated to me that there was really no "sliding scale" of probability related to delay time (i.e., 80% chance of failure if we are delayed just 9 months, 70% chance of failure if we are delayed just 7 months, etc.).

This news also caused me to realize that our original estimate of a 90% chance of failure due to delay was bogus. I now was conscience of the fact that if we received the permits within the next 2 months, the project would not fail due to environmental (really, political) concerns. However, if the permits were not issued within that timeframe, the likelihood of project failure was now 100%.

So, now the question becomes: just how sure am I—expressed as a percent—that granting of the permits would be temporally successful? I did try to get a second meeting with the MEA, but was refused. So, now what?

I convened a meeting—with some participants joining via video conference—of all company employees and some contract personnel who had in the recent past had experience with requesting and being granted permits from this country. After much facilitated discussion, the consensus was that economic pressure on the host government to get the existing plant back on line was relatively intense.

It was suggested that a letter from our company CEO to the country's President, a letter in which the urgency of the issue was conveyed, would cause action to be taken by the MEA. Although this tack is not the best ("going around" the MEA and pulling rank, so to speak), it was determined that there was just too much at stake to let typical protocol get in the way.

Our CEO was summarily contacted and agreed to get such a letter to the President posthaste. With this assurance, it was mutually agreed that the probability of failure due to environmental concerns should be set at no more than 15%. And so it was.

A salient point that I have attempted to make with the story above is that modification of initial estimates can be treacherous. It's relatively easy to take an initial estimate to zero when you believe that the agreed-upon mitigation or capture action (the "fix" for the risk) has been successfully executed and that the threat no longer exists or that the opportunity has been or will be captured. Too often, however, such actions are only partially successful in addressing the risk, or, subsequent experience exposes the originally agreed-to actions to have been naïve, or, at least ineffective.

Subjective methods were used to arrive at the new probability (15%) in our story. If this bugs you, then you likely are not cut out for work in the risk field. When teaching classes on the methods of capturing data for the purpose of building distributions—a means of expressing uncertainty—it has many times been suggested that those data-capture methods "are introducing subjectivity" into the process. My unwavering reply is always: "No, we are just admitting and capturing the subjectivity that already existed." Such subjective methods are ubiquitous and every effort should be made to capture the logic—"on paper"—regarding how and why the quantitative result was deemed reasonable.

Not a Consistent Deciding Body

I'll relate here just one more ramification related to scenarios in which a well-thought-through risk-monetization or risk-management process has not been put in place. I include this short section because I have so often observed this situation in real life.

At least in the U.S., it is built into the corporate fabric the idea that knowledge is power. Ergo, sharing knowledge is a means of lessening one's "stroke." In such a corporate environment, when it comes to generating and sharing the results of a risk assessment regarding a particular corporate division, that assessment rarely is shared across silos. In fact, in such situations, the only time risk reports "see" one another is typically in a single meeting of the heads of departments. In that meeting, only peripheral—and certainly not embarrassing—information is shared.

Such a scenario amounts to having N deciding bodies, N being the number of departments and department heads that have created risk reviews. An up-the-chain corporate executive will swear that he or she has taken account of all the risks because he or she has met with all of the significant department representatives and they have relayed to him or her the individual risk analyses. It is rare, in such meetings, that significant risks that might emanate from a particular discipline will be given the exposure deserved.

Peer reviews sometimes are utilized in an attempt to remedy this affliction. However, I have known such reviews to be mostly procedural ("I won't hurt you if you don't hurt me ..."). Besides that, it is not right thinking to believe that folks in one discipline can adequately or completely review the detailed risk events in another department.

So, if we have N "deciding bodies" making decisions regarding just what and how risks are to be related to decision makers, then my opinion is that such a process is definitely better than having no risk assessments at all, but falls far short of the amazing potential that properly integrated (and not filtered) risks assessments can yield.

Rewarding Activity

This section loosely relates to the section "Organizations That See the Benefit of Doing This—The Incentive System" of Chapter 1. However, a different aspect of the reward system is addressed here.

Stop me if you heard this one. Two guys are driving down the road in a car. Suddenly, the car emits a horrible noise and quits running. The driver knows that he has had a faulty transmission and also knows that it has finally died.

They coast to the shoulder of the road, and the driver gets out, opens the trunk, and starts to change one of the tires. The passenger is perplexed and says, "Joe, why are you changing the tire? It's not flat and it's the transmission that is shot." The driver replies, "Yes, I know the tire is not flat and that the problem is the transmission, but I don't know how to fix a transmission and I do know how to change a tire—look how busy I am!"

This bit of silliness relates to the fact that in many companies, people are rewarded for activity. If you processed 100 forms—regardless of the type of form—then that would be more highly regarded than if you had processed only 50 forms. And so it can be with risk mitigation.

Especially when increased value is your aim, one of the things to guard against is the propensity of individuals to set about mitigating risks for the sake of "knocking off another one." When the incentive system rewards activity, this type of behavior can be incredibly difficult to curtail.

It might seem as though this type of responsibility is outside the remit of the RPP, but the proponent should have, before starting any risk process, a fairly clear idea of what motivates the individuals in any discipline or company segment that will contribute to the risk process. If any part of the proposed risk methodology flies in the face of the behavior for which a person gets rewarded, then it is up to the RPP to be sure that when he or she proposes the risk-process plan, recommended activities and responsibilities do not directly conflict with the entrenched incentive system. It is a bit difficult to avoid adding some work to the everyday workflow for individuals, and it is up to top management to convince department personnel that the added work will more than pay off in the form of increased efficiency, more profitable and successful projects, and, therefore, job security. This argument, however, will not convince someone to "buck the incentive system."

For example, you might propose that as part of the risk process, an individual travel once a month to another company site to compare notes with a peer. The individual, for whom travel is recommended, however, might be getting rewarded for travel dollars saved. That is, part of the individual's remuneration is a "cut" of the travel dollars not spent relative to some benchmark. Though the individual might be convinced that the risk process is a good thing, in this case, they are still not likely to accept a reduction in their personal pay check just because you recommended they travel as part of the risk process. As part of your risk process, you have recommended an action that runs counter to the incentive system. Don't do that!

Suggested Readings

The list below includes material that the reader might find helpful and relevant to the subject matter discussed in this section of the book.

Newendorp, P. D., and J. R. Schuyler. *Decision Analysis for Petroleum Exploration.* 2nd ed. Aurora, CO: Planning Press, 2000.

Saltelli, A., K. Chan, and E. M. Scott. *Sensitivity Analysis.* New York: John Wiley & Sons, 2000.

Saltelli, A., S. Tarantola, F. Campolongo, and M. Ratto. *Sensitivity Analysis in Practice.* New York: John Wiley & Sons, 2004.

Vose, D. *Risk Analysis: A Quantitative Guide.* 2nd ed. Chichester, UK: John Wiley & Sons, 2000.

6

One Route to Risk Monetization—The
"Perfect World" or "High Control" Path

It's an imperfect world. I hope that doesn't shock you. The imperfection of the real world does not, however, keep we humans from dreaming at night— or during the day—about the "perfect" somebody or situation that would make us happy. This is the Pollyanna approach I will take in this chapter regarding risk monetization.

When most people hear the term "monetization," their gray matter conjures up images of quantitative processes and analysis. While such visions are not completely without merit, it is undeniable that the technical and quantitative facets of the monetization process take a back seat to the cultural, political, and organizational aspects. Advice in this chapter will be aimed at both the project manager (PM) and at the risk-process proponent (RPP).

Try to Be Unfashionably Early

To be successful at risk-monetization-process implementation, it is important, but not absolutely critical, to intervene as early as is practical in the cultural, political, and organizational processes already in place in the project group. If you are lucky enough to be able to be part of a project team when it is initially forming, so much the better. However, in my experience, that is rarely the case.

Most typical is that someone initially has an idea for a project. That project, before it can be sanctioned (i.e., officially begun by being staffed and budgeted), must make a proposal to decision makers. Necessary to the proposal is the involvement of technical, financial, commercial, legal, environmental, logistical, health and safety, and other experts. At this point, the project has no official budget and is funded out of a "general fund" and is considered as part of the "cost of doing business."

Individuals lend their expertise to the creation of a "base-case" scenario. That scenario typically is captured as a model in a spreadsheet (or multiple spreadsheets). Technical and other information and data are input to the model and the result is usually, ultimately, some measure of value such as net present value (NPV), internal rate of return (IRR), or another discounted-cash-flow-type metric. Regardless of whether NPV, IRR, or some other measure is calculated, the generated value is utilized to compare the proposed project to other investment opportunities. If the project is deemed to be worthy of pursuit, then it is officially sanctioned and budget and staff are allocated.

A fundamental flaw in the scheme outlined above is that, for the most part, the quantitative metric utilized to represent the perceived value of the project (NPV, IRR, or whatever) rarely if ever accurately reflects the impact of a comprehensive and quantified set of threats or opportunities (i.e., risks). Having said that, I have to back-peddle just a bit regarding that statement because it is not unusual for the "base-case" scenario for a project to incorporate most opportunities (things that could happen that would enhance the value or likelihood of success of the project)—even if the probabilities of those opportunities coming to pass are relatively small.

Remember, this project usually is in competition with a host of other investment opportunities. There is only so much money to go around, and, therefore, it behooves the project-proposing team to present as "rosy" a picture as is practical. So, the base-case scenario presented for the project can be laden with the consequences of opportunities and bereft of the impacts of most threats. In my humble opinion, comparison of projects that have been evaluated in such a manner is no way to run a railroad, so to speak.

So, what to do? As I stated at the beginning of this chapter, it is an imperfect world. In the perfect world, the RPP would be "in on the ground floor" with regard to project development. That is, he/she would be an integral part of the initially formed team that put together the base-case scenario used to represent the perceived value of the project. If early intervention by the RPP was typical, then the perceived value of projects would better reflect the real-world value of those efforts because low-probability opportunities would be properly handled and relatively high-probability threats would impact value.

If this is such a good idea, then why does it not happen regularly? There are two main reasons. One is that the RPP is not typically seen as an essential part of most projects. You have to have a commercial analyst because you need to know if the project is commercially viable (evaluate the market). A financial expert is necessary because the project team needs to know whether or not funding is available and just how much capital it will take to launch the project. Input from technical experts is necessary because project members have to propose a project that is technically feasible and practical. But, do they really think they need a person to, in their perception, throw cold water on the process (the RPP)? Usually not.

The second main reason why an RPP is not part of the initial cadre of project experts for Project A is because Project B does not employ an RPP. If it is typical, and it is a base-case scenario for a proposal reflects the capture of more opportunities (than is warranted) and does not reflect most threats (that are warranted), then why would Project Team A wish to employ someone who will reign-in the impact of opportunities and better reflect the consequences of threats—especially when Project Team B is not doing so?

There is no question that better decisions regarding the value of projects could be made if those projects initially employed an RPP. However, the real-world situation fundamentally precludes RPP participation in the initial stages of project design and valuation. This is not right, but that's the way it usually is.

If the projects in a company's portfolio are those of short duration—say, a year or so from inception to completion—then the risk process has some hope of being recognized as something that should be initially embraced. Hope comes from the fact that when dealing with a relatively large number of short-term projects, it is possible to compare the ultimate value of the project with the value that was initially proposed for consideration for sanction. Consistent optimism in the value of the base case relative to the real-world result can be ammunition for the argument that risk processes such as monetization would yield initial estimates of value that are more in line with reality. Inception of a risk process can, in the case of a relatively great number of short-term projects, demonstrate the positive impact of early risk evaluation.

In organizations that mainly deal with a relatively small number of long-term projects, it is much more arduous a task to demonstrate the positive impact of early in the project risk intervention. Such projects simply take so long and pass through so many stages and groups that the "paper trail," so to speak, is lost. That is, the deviation in ultimate value from the initially proposed value can, and is, influenced by so many global and local factors over long periods of time that the ability to reflect the impact of a risk process is tenuous at best. In cases like this, one must depend upon the logic that it just makes sense to attempt to capture opportunities and mitigate threats for a better outcome.

Like a broken record (reference for those of you who are old enough—like I am—to comprehend that such a thing), I'll repeat my mantra regarding what is important. While later in this chapter I will detail the quantitative aspects of risk monetization, I just can't stress enough that all effort on the risk assessment/management/monetization front is wasted energy if the cultural, political, and organizational facets have not been successfully addressed. As documented immediately above, early intervention is best.

Culture

If you are a PM who toils in a culture that disseminates volumes of rules and regulations but in practice tolerates a "wink and nod" approach to those edicts, then I would suggest that your time might be better spent in pursuit of other goals rather than attempting to implement a risk practice. For any risk process to function as intended, the culture within which the risk methods are to be utilized has to be one in which the "cowboy" approach is eschewed. In such a culture, adherence to safety regulations and to ethical and moral standards needs to be taken seriously and actually practiced by all members of top management. Subordinates should emulate that behavior. If you find yourself in a "fly by the seat of your pants" or "just don't get caught" environ-ment, then do yourself a favor and stop reading now—"it ain't gonna work."

Instituting any risk process precipitates two main upheavals of the cul-tural norm. The first is that implementing a risk process does mean at least some additional effort will need to be expended. Second, it means changing the way things currently are done. Neither of these attributes will be conge-nially accepted without the communicated and demonstrated dedication of leadership. If organizational leaders believe in the process and communicate to followers that it is in their best interest to accept the modifications of the routine for the sake of the greater good, then a risk process has a chance of taking hold.

When I say "taking hold," I mean that practitioners will not only faithfully adhere to the risk process when it is "fresh" and new to the organization and while the eye of management is "on them," but the parties effected will continue employing the risk techniques as part of their normal and accepted workflow. Impact of risk-process implementation is typically seen over the "long haul" and is sometimes not as evident in the short run.

Politics

There is no question that this chapter could easily be expanded into a book. In fact, I did write an entire book—*Modern Corporate Risk Management—A Blueprint for Positive Change and Effectiveness* (2007, see Recommended Readings at the end of this chapter)—that focused on the reasons why risk processes are difficult to implement and on solutions to the prob-lems. A cursory review of the cultural facets of risk-process implementa-tion was presented in the previous section. Although significant, cultural peculiarities rank low in their insidious nature relative to the dark world of politics.

I have returned to the theme of "the incentive system" numerous times in this book and will do so again in sections to follow. While the fixation of this book is not the incentive system, it is nonetheless true that the reward structure is at the heart of—or near the heart of—the reasons for the actions that people take.

A thorough understanding of an organization's political landscape might seem peripheral to the subject of risk monetization, but in reality it is central. The links between political actions, the incentive system, and implementa-tion of a risk process are tangible and real.

Even in an amenable culture, political realities exist at multiple levels within an organization. That is, some political pressures emanate from out-side the organization while others are decidedly local.

Just by way of example, it often is the case today that projects are very expensive. One way for any company to reduce capital costs is to collaborate with another company ("partners") or a consortium of companies. Another capital-reducing tack is for a company to act in "partnership" with a host government or government agency. In any case, political realities come to the fore.

Regardless of how well entrenched a risk process might be or how indoctrinated the company personnel, it is regularly the case that company "partners" don't share the propensity for tolerating a risk-based approach to business. A risk-centered proposition requires understanding and utilization of probabilistic data. Cost, revenue, timing, and other aspects of the project might all be probability weighted and, worse yet from the risk-free perspective, represented not as "sure" deterministic values, but by ranges. Nothing drives the traditional "spreadsheet crowd" over the brink faster than lack of a single value to type into that beckoning cell.

If the partner in a venture—especially if that partner is a government or government proxy—has an adverse reaction to dealing with probabilistic and uncertain information, then there will come significant political pressure to abandon the risk-based approach within the company.

It is usually folly to expect or even dream of enticing the partner or partners to embrace a risk-based approach. However, I have always found it interesting to view the reaction of risk-adverse partners to the insights revealed by a well-constructed and meticulously populated risk register (RR). I have actually had the experience of partners asking: "Hey, how can we do that?" It also is bad practice for any company to eschew a probabilistic methodology just because a partner is not enlightened in such matters. My best advice is for the risk-converted company to continue to adhere to risk processes but to realize that information so generated will have to be "translated" prior to being shared with the partner.

This piece of advice, I realize, is more easily said than done. Typical is the scenario where company and partner personnel are in meetings discussing some facet of the project. Company personnel presenting probabilistic data and partner representatives presenting deterministic information is a recipe for confusion and tension—at least. A best defense against the likelihood of such potential turmoil is for company personnel to agree amongst themselves that when interacting with partner representatives they—the company folks—will present preagreed single values for predetermined parameters (mean value for a revenue number, P90 value for a cost, etc.). By no means is this a perfect solution, but it is the best way I have found to preserve a company's commitment to the rigors of a risk process and to foster a harmonious relationship with risk-adverse partners.

A veritable cornucopia of examples of "external" political influence exists regarding utilization of risk practices, but space in this book precludes expansion of this subject in this venue. I urge the reader to contact me directly if more discussion or advice regarding this topic is warranted. Now, a brief expose of internal political influence.

I would first like to point out that the scenario related in the preceding paragraphs is not unique to a "company vs. external-partner" situation. The situation described above can manifest itself within any company.

For example, it might be standard practice for the construction arm of a corporation to embrace the practice of generating risk-based forecasts and plans. However, such predictions and plans might not be welcomed at higher levels in the organization. This can often be the case when an external audit firm has little or no tolerance for probabilistic data. Higher echelons of the corporation are required to provide data to external auditors. If top management are being "fed" probabilistic data from corporate arms but have to deliver deterministic information to external entities, then unless a prearranged and well-understood translation practice is in place, it might be politically untenable for any arm of the company to deliver risk-based information "up the chain."

This is not to say that such risk practices and information can't be utilized to improve the quality and financial relevance of any project in any arm of the company. However, if at the "top" of any chain there exists a risk-averse recipient of the data, then it should be clear that risk-based information needs to be translated in a consistent manner prior to being related.

At the beginning of this entire section, I related to the relationship of politics and the incentive system. Until now, I have not addressed that link. I'll now fix that.

In previous sections of this book, I have described the propensity of organizations to avoid discussing and/or documenting—especially threats— that relate to "sensitive" areas. Such areas might be represented by forecasts of injuries or fatalities and their impacts on financial aspects. Recording such data in "open" RRs typically is not done. In a well-orchestrated risk process, such sensitive data still can be addressed, but a separate "need-to-know-only" risk report can be separately and discretely generated.

Such separately considered risk reports usually are not created because the company does not want to realize and understand the impact of, say, fatalities and injuries. Such reports are generated because our ethical and moral sensitivities implore us to not consider such events in the same manner that we consider other more mundane threats. However, in some organizations—even those that have mainly embraced a risk culture—there can exist a strong political incentive to "not know" or to officially remain ignorant.

To discuss threats or opportunities and to record them is to admit that you know about them. Such discussions or documentation can preclude plausible deniability. Deniability and ignorance, in some organizations, is seen as tantamount to innocence. While ignorance is never an excuse, it can nonetheless be true that the incentive system within a company can promote the adaptation of the "I didn't know" approach.

An incentive system and internal politics can be "hard" on those who knew a threat might materialize but took insufficient or inadequate action to prevent the threat from materializing. That same incentive system can "go easier" on individuals who can claim that the threat "just wasn't on the radar." Such incentive systems—and they are ubiquitous—promote the approach of institutionalized ignorance. Therefore, if you are a person within such a political structure who desires promotion within the existing political realm, it does not behoove you to have conversations or to document the existence of significant threats.

Political situations such as described above—and many others—can be significant hurdles to the implementation of a cogent risk process. The best defense against these impediments is to fully acknowledge and understand "the system." Knowledge is power, and if the purveyor of a risk process is aware of the incentive system and political realities of the organization, then steps can be taken to design a risk process that "fixes" the problems that can be remedied and blends best with those impediments that can't be resolved. Like I said at the beginning of this chapter, it's an imperfect world.

Organization

It might seem that the preceding discussions of culture and politics should have covered most aspects of organization. It is true that consideration of an organization will include cultural and political aspects, but there are more structural facets to be considered.

Central to most organizations is the "organization chart." Such diagrams typically outline reporting relationships. That is, they show who, for example, performs annual reviews of whom. However, such diagrams rarely impart the internal organizational intricacies regarding how decisions are made or how the organization actually "works."

In a "command and control" (CAC) organization, diagrammatic representation of reporting relationships comes closer to documenting "how things work" than a similar graphic in, say, a matrix-type organization (more on the matrix below). It is absolutely so that every organization is a blend of the CAC and matrix approach. In the CAC-type structure, however, it might be true that any part of the organization that desires to implement a risk-based process needs mainly to seek sanction from the organizational entity that is "above" it on the chart.

In a CAC organization, then, permeation of the risk approach can mainly be a matter of the risk process "trickling up" or "trickling down" the in-line organizations in the diagram. As described in the previous section on politics, the migration of a risk process across any organizational boundaries—"in line" on the organization chart or not—can be an exercise characterized by resistance of many sorts and personal trepidation. Risk-process implementation in the CAC-type organization, however, is "a piece of cake" relative to instantiation of such a method in the matrix organization (MO).

Career advancement in a CAC-type company usually includes at least some aspect of being favorably considered by those individuals or groups that are superior to you on the organization chart. While such favor is not irrelevant in an MO, the real power in any MO is gaining a real and practical understanding of how the organization actually makes decisions and gets things done.

In an MO, the organization chart mainly indicates, for example, who performs reviews for whom. Such diagrams do not, however, give much of an indication regarding what entities of the organization interact to effectively make decisions. In an MO, the interplay of various disciplines can be exceedingly complex and can take place at low levels (as opposed to a CAC organization in which decision-making interaction tends to happen at higher levels). So, what does this have to do with risk stuff?

As previously stated, in a CAC-type organization, permeation of a risk process can be a mainly "trickle up" or "trickle down" process regarding the outlined chain of command. In an MO, however, the interplay of organizational entities at low levels can indicate that if a risk process is going to be utilized by one element of the matrix, such a process likely has to be adopted by all (or most) other entities that constitute the decision-making process.

Passing of probabilistic- and range-based data from a risk-process-embracing organizational entity to other disciplines in the MO decision-making mix likely will be the source of internal confusion and inefficiency, at best. In an MO, then, adoption of a risk process has to be more "horizontal" across organizational entities rather than mainly "vertical" (up and down the chain) as in CAC-type organizations. This might at first seem more onerous, however, I have personally discovered that the spread of the "risk religion" in the MO can be less of a chore than in the CAC, and here's why.

In an MO, each in-the-line-of-decision-making entity typically has one or more individuals who are responsible for looking out for threats and opportunities even if that process is not formalized. Any risk-process purveyor should identify those individuals or groups in each organizational entity and convene a meeting with them. With the backing of upper management, the risk-process purveyor can present the well-thought-through plan (and that person had better have one!) of how the risk process will be implemented and how each entity plays an important role.

If the right individuals within the various entities can form a community—one of support and mutual consideration—then spread of the risk philosophy in an MO can be facilitated. Believe me, it is all about making each entity realize their significant role in successful implementation, creating a community, and lending support to those who are less capable so that their resistance to the process is minimal.

Appoint an RPP

In the "Many Views of Risk" section of Chapter 1, in a cursory fashion I outlined the basic attributes that an ideal RPP should possess.

Those characteristics are the following:

- Able to speak intelligently with each—or at least, most—of the personnel in each discipline, that is, be cognizant of the colloquial verbiage.
- Able to understand how personnel in any given discipline view threats and opportunities.
- Able to understand the metrics employed by any discipline.
- Able to understand the expression of the metric employed.
- Able to translate the utilized expression in to a common metric that can be used to impact the perceived value of the project.

In that same section of Chapter 1, I mentioned that finding an individual that is so capable likely is "a tall order." Given that this chapter is concerned with what we would have if we could have it—that is, the ideal world—I will here expand significantly the discussion of what it takes to act as an effective RPP.

The bullet points above are just a cursory outline, and my discussion here will not necessarily address those points in order. However, by the conclusion of this section, each of the bulleted attributes will be addressed.

Not a Jerk

I have previously mentioned that in an organization, a group tends to take on the opinions and behaviors of its leadership. For example, if the group leader takes a "wink and nod" approach to rules and regulations, you can bet that the organization under that leader will adopt a similar attitude toward company dictates.

In this section, I will mainly speak of the RPP role as one that is filled by a single person. However, it is not unusual for the RPP position to be represented by a small group. Just like the leadership example given in the preceding paragraph, the seriousness and professionalism with which the RPP team approaches the requisite tasks will reflect the attitude of the group leader.

I am well aware of the fact that most people who peruse these pages expect a book on risk monetization to be "all about" quantification, equations, and the like. Although such things have their place in the scheme of things and will absolutely be addressed in this chapter and following chapters, I can say with absolute assurance that technical and quantitative issues take a back seat to those aspects that relate to getting people to act in the manner required for risk-monetization implementation.

If the person filling the RPP role lacks the "people skills" required, then the likelihood of experiencing a successful implementation is much diminished. Far be it for me to lecture anyone on how to be a "nice person," but being able to amicably interface with a broad spectrum of discipline experts and others is a "must."

Respect is a two-way street, and success largely hinges on respect. The RPP ideally is someone the organization holds in high regard. This individual (or group) need not be a jack of all trades but should at least garner respect within his/her own area of expertise. Even more importantly, this person ideally should have a reputation for not being a link in the gossip chain—that is, the person should be perceived as a trustworthy confidant. If others believe that what they reveal to the RPP will be shared with any part of the larger community, the ability of the RPP to wrest from personnel sensitive risk-related information will be greatly diminished or rendered nonexistent. This is the second way of the two-way street of respect—the RPP has to have respect for the individuals who share information and for the perceived confidential nature of the information itself.

Not a New Hire and Not a "Good Old Boy"

It often is the case that the RPP position is new to the organization and, therefore, mainly misunderstood. Misconceptions concerning the position extend to management levels.

It is not uncommon to view the RPP slot, at least initially, as ancillary. The "experts" are all busy, and even if they were not, the RPP position would commonly be viewed as "below their station" and, therefore, eschewed. With most qualified candidates at least acting frantic, the RPP position can "trickle down" to a relatively new member of the organization. This is a mistake.

As mentioned in the preceding section, the person who fills the RPP role should command the respect of the organization. It is a rare instance when someone newly hired is so perceived. Respect is earned, and the process of earning takes time.

Although the RPP position will morph into one of the more potent stations on the project team, it is only through careful, clear, and persistent delineation of the RPP responsibilities that the post will attract a willing experienced hand. Such delineation requires that the PM first understand the role and its implications. The scope of the RPP role will be intimately described in the next major section of this chapter.

Just as it is a fundamental blunder to appoint a relatively new-hire as the RPP, you always have to be on the lookout for a senior person on the staff who is seeking a place to reside until retirement. It is not that people in that position can't adequately perform the required RPP tasks, but what you are trying to avoid is being saddled with an RPP who lacks true enthusiasm for the job and who is attempting to procure a position in which minimal effort will be tolerated.

Even if the RPP position is filled by a "go getter," you should avoid appointing as the RPP someone who is a charter member of any "good old boys" club. That is, relatively senior staff members tend to be "chummy" with like members of other disciplines. Anyone appointed to be RPP is going to have to occasionally "crack the whip" and have frank conversations with individuals or groups who are perceived as not being in full compliance. Members of the "good old boys" club rarely pass judgment on one another and will typically find it at least exceedingly uncomfortable to try to bring into line one of the other senior and charter members of the club.

Understand the Decision-Making Process

Well, I spent the previous two subsections talking about what an RPP should not be. I suppose it's time to consider just what that person should be.

First and foremost, whether or not the RPP has a deep and abiding comprehension of the workings of any single organizational entity, it is critical that the chosen one have a practical understanding of how the organization works. When in this context I employ the term "practical," I am more than implying that the RPP's grasp of how the organization makes decisions has to be the actual decision-making process and not one gleaned from an organization chart or other textual or graphic representation of the relationships of the relevant entities.

To possess this type of knowledge, it takes time and exposure to the various tentacles of the organization. This insight can only be gained by being part of the decision-making process over time. This is my book, so like grandpa spinning pointless tales, I'll take a bit of space to relate to you my first encounter with how organizations really work.

I was a lowly graduate-school teaching assistant at a major university in the northeast of the United States. To make a long story short, I needed to be granted approval for a project that involved getting sanction from several departments. I got hold of the university organization chart and deduced who were the heads (chairpersons) of the departments from which I needed permission. I approached the Chairman of my department. He listened to my request, suggested that he would get right on it, and I waited … and waited.

Nothing happened. One day by happenstance I found myself in the office of Bev, our department's Administrative Assistant. Quite inadvertently, I mentioned my frustration with the approval process for my project. The very next day, I had the permission I sought.

Turned out that regardless of what the organizational graphics depicted, it was Bev and her network of friends and peers who truly held the power to get things done—and, more importantly—to actually make decisions. Later, I was convinced that Bev could have gotten the world turned inside-out if she and her connections deemed it necessary. Furthermore, I came to learn, the Chairman of the department also depended on Bev and her connections to get his work done. So much for organization charts.

I think you get the point. It is not so much about knowing which individual to see (like Bev in the story above), but rather it is important to understand who are the actual individuals in the decision-making process matrix and how they relate to one another. This can truly be known only by exposure over time to the organization and its subterranean processes. Such insight is critical for the RPP because the risk assessment/management/monetization process will span multiple arms of the organization and gaining sanction from exactly the right individuals is crucial.

Knowledge of Disciplines

Now, I am not so silly as to expect that any individual would have a deep understanding of each discipline that will contribute to the risk process. Nobody I know is a lawyer, a commercial analyst, a chemical engineer, a security expert, and so on. However, it is not unreasonable to expect the RPP to have a fundamental comprehension of just what role each discipline plays in the execution of a project.

First, it is fundamental that the RPP be aware of the various disciplines. For a typical project, it is not uncommon for the legal, negotiating, finance, commercial, logistics, chemical engineering, mechanical engineering, electrical engineering, IT, health and safety, logistics, security, medical, supply chain, and other departments to be involved. Simple awareness of the scope of the contributing entities is a must.

More than that, though, a cursory understanding of what role each entity will play in the project is something the RPP should possess. That is, the RPP might comprehend, just for example, that the legal department would get involved at the very beginning of the project to pass judgment on whether or not any laws or corporate edicts would be violated in pursuit of the project. Such things as "advance payments," needing to obtain equipment from sanctioned countries, and other project requirements would need to be handled by the legal department.

I am not here suggesting that unless the RPP is aware of the role of the legal department, for example, that the lawyers would not be involved in the project. Of course, they would be. However, as has been stated many times before in this book, the aim of a comprehensive risk-monetization process is to express the impact on perceived project value of threats and opportunities emanating from all pertinent disciplines. The RPP can't—and won't—account for legal risks, for example, if he/she is not aware of the role of the legal department in the overall project process.

So, it is not enough that the RPP have a grip on the decision-making methodology and the entities that comprise that process. It also is critical that the RPP have at least a fundamental understanding of just what role each entity plays in the project lifecycle and to be able to incorporate the risks from those entities in the risk-monetization scheme. This leads me to my next point.

Learn Some Jargon

Nobody appreciates a wise guy. Prior to a meeting with the chemical engineers, for example, you don't want to be perceived as an RPP who says, "Yeah,

I knew I was coming to see you today, so last night I picked up 'Chemical Engineering for Dummies' and now I know all about this ... ." This would be at least one road to ruin.

Conversely, it is not a good idea to show up at the desk of a discipline expert and have absolutely no idea how to talk to that person. That is, having a working knowledge of the fundamentals of that discipline and being able to understand and express basic jargon is a boon to interfacing with the experts.

Just for example, in a former life, I performed risk analyses for the legal department of a major corporation. Before interfacing with the attorneys, I consulted with one of the attorneys in that department and requested a list of terms and definitions that I might need to comprehend prior to discussing cases with others in the department. Learning the meaning of such terms as "summary judgment," "deposition," and others was, I am convinced, fundamental to my successful life with the lawyers. So, learn something about what they do, but not so much as to come across as someone who "knows all about this" even though you have no formal education in the area. Be humble.

Understand How Each Discipline Views and Expresses Risk

"All politics is local" said the late Tip O'Neill, former Speaker of the House (of Representatives). So it is with risk. Unless all disciplines have been subject to a mass conversion and confluence of view brought about by, for example, a common risk process imposed by the corporation or main organizational entity, each organizational cul-de-sac will embrace a unique view of risk. To exemplify this, I always fall back on the same example—that of the difference in views held by the employees of a finance department and those held by health and safety personnel.

In the finance area, risk is seen, at least in large part, as a good thing. Low risk, low reward. Higher risk, higher return. These folks are seeking risk—not more than they can handle, of course—because it is their job to maximize the return on investment. Contrast that view with the perception by health and safety personnel. Risks in this area are mainly seen as threats. Risks are to be, according to the health and safety view, identified and eradicated.

Different conceptual views are not the end of it, however. For example, it is not unusual for a PM to be equally concerned about time (schedule) as he/she is about money (budget). Contrast that with the construction engineer who fundamentally is focused on the chance of failure of the type of metal to be used in an application or on the probability of successful implementation of a complex machinery design. Security experts will be obsessed with the risks associated with breaching a facility perimeter and avoiding being involved in the local conflicts, while logistics experts worry about the practicality of various transportation options.

It has always been my contention that it is folly to expect that you can practically get all parties to view and express risk (threats and opportunities) in a similar manner. I hold that you would not want to do this even if you could. So, changing views of risk on the "front end" is a waste of time. As long as what they are doing does not violate any laws of the universe, just let them do what they are comfortable doing.

This, of course, puts the burden on someone like the RPP to bring all of these disparate views and expressions of risk together, so they can be integrated and monetized. This no small feat will be addressed later in this book, but suffice it to say here that any credible RPP absolutely has to sport at least two traits.

One is that he/she has to have a tolerance for diversity. Just like being a "people person," the RPP has to take them as they come. Being able to appreciate the colloquial view and expression of risk held by various organizational entities is a necessary trait. The second characteristic that any good-at-their-job RPP must possess is that of being capable of truly understanding and integrating the broad spectrum of risk expressions presented.

For example, legal analysts are likely to present prose as their expression of risk. It is common for engineers to create probabilistic plots and charts to express risk. Those organizational entities that are mainly concerned with risks to the schedule are likely to present gantt charts or some other time-based graphic. And so it goes for each discipline. Believe me, it is no small feat to be able to bring together these various expressions of units, metrics, and plots/charts to create a "unified field theory" of risk. An example of at least one way to accomplish this task will be the subject of a section later in this book.

Suffice it to say that if you can identify a person (or small group) who embodies all of the aforementioned attributes, you should consider yourself exceedingly fortunate. Such individuals do exist, but are generally sought after by other parts of the organization and are difficult to "shake loose" for service as an RPP. For this reason and others, it is not unusual for the RPP position to be filled by a small group of individuals each of whom possesses some subset of the requisite attributes.

Have "The Talk"

In this section, I will be speaking both to the PM and to the RPP. All of the advice given in this section should be implemented as early in the project life-cycle as is practical and possible. I know from personal experience that if the PM and RPP do not "come together" on all of the points listed below, then the probability of experiencing a successful risk process is greatly diminished. So, early in the project life, the PM and RPP need to "have the talk" and document just how their respective roles will be utilized in the execution of the project and what "powers" are assigned to each. This is absolutely essential.

Establish the Powers Bequeathed to the RPP

It has been my experience that, at least at first, the role of the RPP is viewed by project management as an "add on" position. It is a role that is ancillary to the actual work done on the project. Viewing the RPP position such as you would an auditor, for example, is misguided.

Establishing "up front" that the RPP role will have real power and that some of that power will be usurped from the PM position is necessary. The PM needs to view this relationship as a delegation of responsibilities and not as a diminution of the PM role. One of the last things you want is to have "the parents fighting in front of the kids." That is, it is not particularly helpful if the PM and RPP are in regular disagreement about project issues. Clear lines of demarcation have to be established, and the RPP position needs to be perceived as one of real authority. Just a subset of the areas of real impact are documented below.

Can Call Off Critical Meetings

Early in most risk processes, threats, and opportunities are identified and documented. There exist myriad ways to do this, but one of the tried-and-true processes is to convene a risk-identification workshop (RIW). A detailed rendering of an RIW will be presented later in this chapter, but allow me to jump ahead a bit here.

It is the responsibility of the RPP to identify individuals whose attendance at the RIW is critical. Such personnel are those who possess the requisite expertise in a particular field—that field being a critical aspect of the project. For the RIW and subsequent integration of risks to be successful, expressions of threats and opportunities from knowledgeable staff from each critical discipline is required. Without true and expert insight into how a given discipline operates and how that discipline might interface with other critical project elements, effective identification and description/documentation of risks is not likely.

For example, it is not sufficient for a junior chemical engineer—one with little project experience—to represent chemical engineering at the RIW. Only a seasoned representative might realize that the chemical process to be employed in the project will require, in the particular project context, special permits to be issued. The experienced engineer will be cognizant that early interfacing with representatives from the legal and environmental departments is necessary if permits are to be issued in a timely manner. One of the threats, therefore, that might be identified by the experienced engineering professional is that the legal and environmental aspects of the chemical process might not be accomplished in the timeframe or budget required.

Any invitations to an RIW should go out to select staff under the signatures of both the PM and the RPP. It should be communicated to senior staff, well in advance of the RIW event, the essential nature of the RIW and the criticality of their attendance.

Well, even if all that is done and acknowledgment in writing (usually in the form of an e-mail, these days) has been received from personnel whose attendance is required, it has been my experience that—come the day of the RIW—those who indicated that they would attend are either "too busy" to show up, or, they find it convenient to send a less experienced proxy. It should be agreed between the PM and the RPP that if, in the RPP's opinion, the RIW has less than the requisite representation, the RPP has the power to cancel the meeting.

Just like building a machine, if you have requested the parts necessary but substandard or wrong parts are delivered, it most certainly is warranted that you refuse to build the machine rather than construct a mechanism that is doomed to fail. So, it is with the RIW. Attempting to gain the necessary insights from either absent representatives or from substitutes who do not possess the knowledge required is a waste of time. Worst yet, if the RPP allows the meeting to proceed and critical risks are missed or misjudged, the project will suffer and the entire risk process will be viewed with distain.

If the initial attempt at an RIW is canceled, then following cancellation, it should be made clear why the RIW was terminated (list who did not show up) and the damage to the project schedule and budget that cancellation has caused. This should be communicated to all invited parties and their management.

Typically, it is peer pressure and not management edicts that remedy the attendance problem. Senior staff who did show up for the initial but cancelled RIW had to arrange their schedules accordingly. Now, they have to do it again. They have to do it again because John Doe and others did not feel compelled to attend. Communication between those who did show up and those who did not usually does the trick. These are the kinds of little insights that only some guy like me—who has been beaten about the face and head by this process—can have and relate.

A point to be made here is that the PM and the RPP have to be in agreement regarding the RPP's power to cancel meetings. The PM should know full well that cancellation of the meeting and rescheduling of another will have serious consequences for budget and schedule. Such impacts should not come as a surprise to the PM.

Reject Unqualified Representatives

This aspect is related to the section above, but it is different enough to warrant separate attention. I have already related that the RPP should have the power to call off an RIW if less-than-qualified (substitutes for critical individuals) representatives show up at the RIW. However, the potential contribution of less-qualified individuals might be posed well in advance of the meeting.

It can often be the case that the RPP requests the participation of "John Doe," the expert in the discipline. Discipline management, however, are not thrilled about this whole risk process and are even less enthusiastic concerning tying up John Doe in a multiday RIW. So, management offers a less-qualified substitute for John.

Agreement between the PM and the RPP regarding the RPP's power to accept/reject participants needs to be established early in the process. If the RPP rejects the offer of the less-qualified substitute, you can bet that discipline management (who offered the proxy for John Doe) will be, upon receiving the rejection notice from the RPP, standing in the PM's office voicing their displeasure. If the PM is not of a mind to back up the decisions made by the RPP, then it will be quickly known by the entire organization that the RPP is a toothless police dog. Backing by the PM of the RPP decisions has to be established and real.

Control of Data Entry

A typical project can represent the confluence of a dozen or more major disciplines. Each discipline can view its data and other information as sacrosanct and not to be viewed or manipulated by those outside the area of expertise. This attitude will have to be, at least in part, disbanded if the RPP is to successfully manage the RR (see Chapter 2 of this book).

Early on, it should be agreed between the PM and the RPP that it will be part of the remit of the RPP to control entry of risk data into and out of the RR. If it is the RPP's responsibility to manage the risks in the RR and to be accountable for assuring the PM and others that threats and opportunities are being handled, then it would make the RPP's job impossible if flow of information in and out of the RR was not funneled through the RPP.

As I indicated above, it is common that a dozen or more major disciplines are involved in a project. Each of those disciplines might have identified tens of threats and opportunities. Many of the threats and opportunities in the RR will impact multiple disciplines. That is, mitigating a threat, for example, might involve efforts from both the legal and environmental groups. Therefore, the RR will initially be comprised of possibly a hundred or more major risks. Each of these risks can have associated with it mitigation or capture actions that can be multidisciplinary and complex.

It stands to reason, then, that if just anyone is able to access the RR and change the data and other information, such a system would render the RPP helpless in trying to keep up with changes—much less to be able to pass judgment on whether or not those changes were valid. Ergo, it needs to be agreed that changes to the universal project RR can only be enacted by the RPP or an RPP-designated representative. The issue of "how many risk registers" will be allowed is addressed in detail in a following major section of this chapter.

I can tell you from experience that disciplines won't like this idea. However, if chaos is to be avoided with regard to data/information flow into and out of the RR, then some acceptable scheme that allows the RPP to fulfill his/her role of assurance has to be established. Typically, and most effectively, this is handled by the RPP (or a designated representative) having exclusive control of the RR data/information.

More insidious than simply keeping track of things is the propensity for those charged with execution of threat-mitigation or opportunity-capture actions to enter in the RR the successful completion of those tasks when, in reality, they are partially completed or have not at all been addressed. Many people mean to get around to the task and would justify a positive entry in the RR as a promise to get to completing the task. This cannot be allowed.

Putting aside the issue of good intentions, some actions might be "completed," but they might not be the actions that were originally agreed upon. For example, it could have been agreed in the RIW that a threat-mitigation action would be the replacement of a critical valve and the piping that leads into and out of the valve. The responsible engineer might have replaced the valve, but not yet the piping. This partial completion of the task cannot be allowed to be entered as a completed risk-mitigation action. The RPP has to have the authority to determine whether or not complete actions have been taken.

Again, complaints to the PM will be made when the RPP refuses to accept as a completed action something that is less than what was intended. To avoid the toothless police dog syndrome, the PM has to stand behind the RPP's judgment.

Decide for Whom the RPP Really Is Working

This is a critical issue. The first section of Chapter 3 of this book ("Decide for Whom You Are (Really) Working") gives an example of this problem. In that section, I relate a story about a fictitious company called Mineitall and the problems related to establishing for whom you are making decisions. I will not here repeat the essence of that story, but the reader is directed to that section of Chapter 3, to get the gist of what I will here relate.

RPP's can be put in a tough spot. It could be true that advancement of a particular project or an aspect of a project could damage or set back the best interest of the greater organization. Just for example, imagine a case in which a PM is considering a project in Country A. Product from this project would compete internally with Project B (another project inside the company), which is farther along in development in another country. Market analysis indicates that there is not enough global demand for the product to warrant production from Project A and Project B. However, as is typical, the PM for Project A is rewarded for successfully launching the project (I have coined the phrase: "We often are rewarded for successfully launching a project, but not for launching a successful project") and he/she has enlisted you—the RPP—to help do so.

As the RPP, where does your loyalty lie? You might think that corporate management should make the decision regarding the fact that if Project A and Project B are both sanctioned, then there would be a noncommercial oversupply of product relative to market size. You would be right in that assumption, but such decisions often are made only after both Project A and Project B have well-developed—independent of one another, of course—base-case proposals for management to consider. To get to that stage, much time, money, and effort can be, and many times are, needlessly expended.

One of the agreements, therefore, between the PM and the RPP should be that the RPP will work diligently in the service of the project as long as both parties—the PM and RPP—agree that the project is in the best interest of the company overall. It should be part of the RPP's remit—and that of anyone else with a strategic view—to bring to the attention of the PM the impracticality of the project or any part of the project if that effort is deemed to be a detriment to the greater strategy and benefit of the larger organization. I harp on this because I have several times actually found myself in exactly the position where I recognized that local incentives to pursue something ran counter to larger strategic goals.

Before you think that the PM in the story above is trying to put his/her interests ahead of those of the greater organization, let me assure you that such positioning often is not the case. Sometimes, PMs have been in an area of the world, or, in service of the same technology, for long periods of time. Their "world" becomes that technology and that part of the actual world. These individuals are real experts in their limited area. It can be the case that they are not aware of more global developments or the greater strategic plan.

Contrast that relatively cloistered perception with that of the typical RPP. Most RPPs, in their career, go from project to project. Those projects can represent a broad spectrum of efforts. Therefore, it can be so that the RPP has a much better understanding of the global picture than does the PM who has asked to enlist the RPP's help.

In addition, it is part of the RPP's job to attempt to bring together the impacts of risks from a very broad spectrum of disciplines (legal, logistical, commercial, technical, political, health and safety, financial, security, etc.). The nature of this effort naturally affords the RPP a broad view of the business. This can be a view that is not available to a PM that has worked for an extended period on the same types of projects in a given part of the world.

So, although it can be an uncomfortable conversation to have, the PM and the RPP have to come to agreement regarding whose best interest the project is going to serve. It has to be agreed that if the RPP deems that the project or parts of the project will work counter to the greater good, then at least a frank conversation will be initiated.

Don't Underestimate the Upheaval

In the "Am I Wasting My Time" section of Chapter 1 and in the "Be Certain of the Right Level of Support" section of Chapter 3, I briefly touch on the subject of the organizational upheaval that will necessarily result from implementation of a proper risk-assessment/management/monetization plan. Although not all strictly relevant to the subject of this subsection and at the risk of absolutely being redundant, I believe that a proper introduction to this section warrants quoting a small bit of text from the aforementioned section of Chapter 3. In that section, I say:

> Risk stuff is hard because it requires people to do things differently. Risk stuff is hard because it embraces subjectivity. Risk stuff is hard because it cuts across silos and requires the participation and cooperation of seemingly disparate disciplines. Risk stuff is hard because it often runs counter to the incentive systems. Risk stuff is hard because it often offers a range of "answers" rather than "the number." Risk stuff is hard because those at the top of the house often proclaim that they are in favor of implementing risk processes, but only mean that if it all happens "below" them and that their world does not change. Risk stuff is hard because if forces quantification and documentation of otherwise "assumed" aspects. Risk stuff is hard for a whole host of other equally important and valid reasons.

All of these things are so. However, there's even more to the story.

Risk Centric Rather than "Project Centric"

As I said in the Introduction of this text, I strongly suspect that most readers will not plow through this volume as they would a novel—reading from first page to last. Rather, most "readers" of this book will use it as a reference book, typically looking up a particular subject in the Index and perusing only that part of the text that is immediately relevant to the problem of the moment. That's fine and expected.

However, if you are one of those rare individuals who has read from the first page to this point, it should be clear that if you pursue implementation of a risk process in a project, the risk process supplants the typical "project workflow" with a overriding work process that is risk centric. Although a necessary conversion, I can tell you from personal experience that project team members "ain't gonna like it."

First, what does it mean to become "risk centric"? Consider just a part of the typical project—the initial commercial effort. The insipid pattern is for the commercial analyst to consider input from other disciplines such as logistics, market research, and myriad others and to construct an initial model that, in the end, calculates a metric that represents the perceived value of the project. This process, aside from other roles filled by the commercial analyst, requires considerable knowledge and skill.

Now consider the same task performed under the overarching umbrella of a risk process. Suddenly, the information gathered from other sources, such as the logistics group and the market research group, is likely not deterministic in nature (i.e., single valued) but is represented by ranges. As if ranges were not enough, many ranges for parameters would carry with them an associated chance of failure (COF) value indicating whether or not the range would be considered for inclusion in the model.

Values at time periods in the time-series analysis no longer represent single values, but now are depicted as bars representing ranges. Multiple output values such as the expected value of success and the expected value for the portfolio should be generated. Means, P10s, P90s, medians, and other metrics need to be considered. On output plots, COF is combined with the X-axis range to create probability-weighted outcomes. Uhhhhh … ..!!!

Furthermore, when a proper risk process is instantiated, the focus of the group necessarily becomes the RR. Along with well-established tasks, taking threat-mitigation and/or opportunity-capture actions are paramount to maximizing the chance of success of the project and its value. Learning to make the RR one of the centerpieces of the everyday work process is a learned skill.

This represents a lot of change, and this is just an example. In this example, I noted that the logistics and market research groups (among many others) would be feeding probabilistic information to the commercial analyst. Well, how do you suppose the logistics and market research folks arrived at those probabilistic outputs? You guessed it—by having implemented risk-based processes in their areas. And so it might go throughout the entire organization.

Now, if I were you, I'd be thinking, "Glenn, implementing such a great disturbance in the way people work is required, has this ever been successfully implemented?" The answer is: Absolutely, yes.

Success depends on starting small. Begin risk implementation with a relatively small but receptive group. Let's say, just for example, the logistics group. Such groups make projections regarding costs associated with, usually, moving product or equipment from Point A to Point B on the globe (this explanation understates their mission, but please allow me the benefit of brevity). Applying basic risk principles to the task of creating projections—including creation of a RR and implementing all the steps that such implementation entails—can significantly improve the comprehensive nature and accuracy of forecasts. As I have stated previously in this book, nothing sells like success. When top management recognizes the achievements of the risk-embracing group, others will be anxious to follow. As someone once said, you eat the elephant one bite at a time.

New Sheriff in Town

In previous sections of this chapter, I outlined just some of the "powers" that need to be bequeathed to an RPP if that position is to be effective. Typically, project team members look to the PM and more immediate discipline-related management for direction. Implementation of a risk process necessitates that the RPP play a central role in guiding the entire project team. This concept can be very difficult for some people to swallow.

Now, I'm guessing that you think that the project team members at "the pointy end of the spear"—those who actually do the work—would be most opposed to taking direction from an RPP who is not an expert in their field. While it is true that the RPP is not omnipotent, it is not his/her job to impose discipline-specific insights, but rather to assure management that major risks have been identified and that threat-mitigation and opportunity-capture actions—described by discipline experts at the RIW—are being completed in a timely manner.

When an RPP operates within these bounds and does not attempt to parcel out discipline-specific advice, then project team members who are to carry out the identified mitigation/capture tasks generally present little resistance. Real push back, however, can emanate from the management ranks that lie between those project team members and the PM.

It has been my experience that people who are trying to get the job done just want to get the job done (profound, I know!). If a cogent and valid process is laid out and such a process does not violate any dearly held principles, then that's how they will perform the task. However, those who direct such people often can feel as though their expertise and relevance is being usurped and/or diminished. This can be a real problem. I have had success in resolving this potential issue using two approaches.

One approach is to meet with these individuals in one group meeting if possible and to attempt to convince them to embrace the risk process (nothing sells like success!!!) using relevant examples. Be prepared for a long meeting. Above all, listen to what they have to say and have patience. If you are the type of person with a "short fuse," then find someone else to conduct the meeting. In the end, however, these individuals have to be convinced that embracing the risk process is in their best interest (and this has to be true) and that at least some direct interaction with the RPP by their reporting personnel will be necessary.

Another approach which is not mutually exclusive with regard to the one described above, is to relay most information to project team members through their direct management. For example, if a threat-mitigation action is reported to the RPP to be completed but the RPP deems that the action does not honor the task originally described in the RR, the RPP can refuse to "mark as done" the action in question. Usually, this notice of inadequate response and a discussion of just what is required would be a conversation had between the RPP and the responsible project team member.

An alternate plan is to initially inform the project team member's immediate manager of the unacceptable action and hope that the correct message is conveyed to the team member. This, however, is a bit akin to that parlor game in which a short story is conveyed (whispered in their ear) from person to person until, in the end, the final whispered story and the original story are compared—and found to have little resemblance. A twist on this tactic is to convene a meeting with the RPP, the direct manager, and the project team member. This works as long as the PM takes an even-handed approach to the matter and does not come to the meeting with the intent to defend the team member.

There's nothing easy about dealing with human beings who are laboring under an incentive scheme that is not necessarily aligned with the goals of the risk process. Sometimes you just have to know with whom you are dealing and select a method that best suits.

There is at least one more aspect to the "New Sheriff in Town" situation and it has to do with the PM and his/her reporting relationship with upper management. Given that the RPP is solely responsible for assuring decision makers that the raft of threats and opportunities have been diligently addressed, it is his/her responsibility to convey the sometimes complex risk picture to those decision makers. If there are multiple groups of decision makers to which information must be conveyed, then the RPP should be a presenter to each.

It is just absolutely silly to believe that the RPP could relate to the PM all of the intricacies related to the broad spectrum of risks that constitute the everyday palate of tasks with which the RPP has to wrestle. At each major meeting with decision makers, a portion of that meeting has to be set aside to address risk issues and the RPP has to be the person to present that information.

A secure PM will have no problem with this. However, those individuals not secure in their station or those who do not share the limelight well, will have a problem with the RPP presenting to upper management. This is one of the issues that needs to be addressed "up front" with the PM and agreement reached. It simply will not do to have all risk information delivered by proxy.

Decide How Sensitive "Soft" Risks Will Be Handled

This topic has been addressed in a cursory manner in the "Holistic Set of Integrated Risks" section of Chapter 1, in a broad-spectrum way in the "Health & Safety" section of Chapter 4, and in some detail in the "Ignoring Soft Risks" section of Chapter 5. I will attempt here not to repeat much of that information and refer the reader to the aforementioned sections. However, in most of those sections I relate the problems associated with "soft" risks. In this chapter, I am attempting to relate the "perfect world" situation in which we could have what we deem necessary to be successful. So, I will relatively briefly address how "soft" risks would ideally be handled.

First, let me reiterate just what is a "soft" risk. The definition is a bit fuzzy, I'll admit. By my definition, a "soft" risk is a threat or opportunity that is not quantified in the typical risk analysis (base-case-representing spreadsheet) and/or is not represented as a specific line item in such models. Having defined the term, I will now quit encasing the term in quotes.

An example might be cultural risks. In some countries, cultural or religious practices differ from our own. Taking account of those deviations from what we might consider the norm—regardless of from what culture you are starting in—is absolutely critical if you are striving for the best chance of success. However, when is the last time you saw a commercial-analysis spreadsheet with a line item labeled "praying" or "eating habits" with values filled in for each time period? Organizational, political, health and safety, and other types of risks typically can be considered soft by the definition above. So, in the perfect world, what would we do about soft risks?

Recognize Them

In the major subsection "H&S Probabilities and Black Swans" of the major "Health & Safety" section of Chapter 4, I address the issue of recording (documenting) a risk as being equivalent to admitting you have a problem—in the case of a threat. Legal and other ramifications of recording risks are addressed in that section.

In the perfect world, the RPP and PM and possibly higher management need to come to agreement regarding how sensitive information in the form of the description of a risk will be handled. I will not here replow the ground of previous chapters in which I expound upon the practical problems related to dealing with sensitive issues such as fatalities. However, I will here reiterate and expand upon some of the real-world approaches to handling such soft risks.

If legal staff has advised that written documentation of some subjects is disallowed, then it should be agreed that discussion of the attendant risks should constitute a part of every risk review. This needs to be agreed upon by all parties. Such risks should under no circumstances be ignored because of their sensitive nature.

Other soft risks should be handled as separate risk assessments and presentations. These risks fall into two categories—those that are "sensitive" in nature such as injury or fatality assessments and those that represent low probability but high impact.

Most risk analyses culminate with a measure of value such as NPV or IRR, and so on. It is not uncommon to exclude from this type of "cold" analysis the consideration of sensitive risks related to fatalities, injuries, environmental damage, and the like. Such exclusion, however, does not excuse consideration of these risks.

Deal with Them

A typical approach to dealing with such risks is to create a separate risk analysis that will be shared only with those who "need to know." In addition, the costs generated by these segregated assessments are not technically

integrated with the output from assessment of more mundane considerations. Results from the general risk analysis and from the separate "sensitive risk" assessment can be delivered to decision makers as two unique
analyses and decision makers can conceptually "integrate" the results. This
is something that has to be agreed upon "up front" by the RPP and all relevant parties.

A second category of risks to be considered uniquely are those that represent low probability but high impact. The sort of ridiculous example often
used is the commercial airplane falling on the factory. If it happened, would
it be devastating? Absolutely. Is there a very high chance that it will happen?
Absolutely not.

I can't tell you the number of times I have seen these "one in a million"
(and I use that phrase not in a quantitative sense!) risks included in a Monte
Carlo model that was only going to be run for, say, 5,000 iterations. What
nonsense!

Given that we can't today practically integrate such risks with those with
much higher associated probabilities (of happening), what can we do? In the
"Use of the Boston Square" section of Chapter 5, in which I rail against the
use of the Boston Square, I mention that is this section of the book, I will
demonstrate an alternative graphical method for presenting risk data. So I
shall.

In that same section of Chapter 5, I make the case that it matters whether
you are with an action attempting to mitigate—in the case of a threat, for
example—the probability of the threat materializing, or, the impact the
threat will have if it does materialize. I use a story about car manufacturers
and airplane manufacturers to illustrate the point.

On a Boston Square, because the Y-axis represents the product of probability and impact, it is impossible to determine which attribute is being
addressed by the X-axis position—the X-axis typically representing the ability to mitigate (or capture in the case of an opportunity) the risk. I allude to
the use of probability/impact plots to remedy the situation.

Figure 6.1 shows a probability/impact plot. In this graphic, the probability
is plotted along the Y-axis and the impact along the X-axis. In such plots,
low-probability/high-impact threats and opportunities are clearly and properly positioned. On the Boston Square, because a low probability multiplied
by a high impact will result in a "middle-of-the-road" Y-axis value, such
risks are indistinguishable from those having truly middling probabilities
and impacts.

As I illustrated with the car/airplane example in Chapter 5, it matters
greatly whether your mitigation action, in the case of a threat for example, is
aimed at affecting the probability or reducing the impact. Figure 6.2 shows
two plots—a probability versus ability-to-mitigate/capture plot and an
impact versus ability-to-mitigate/capture plot. In these diagrams, it is easy
to illustrate which, if not both, attributes are targeted for mitigation. Note
that threats and opportunities can be mixed in the same plot.

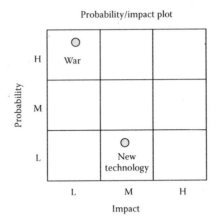

FIGURE 6.1
A probability/impact plot.

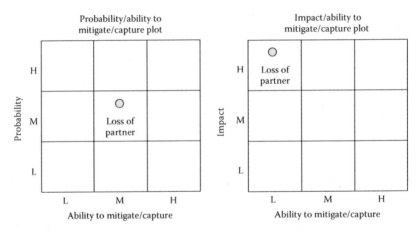

Plot (left) showing that there is a medium probability of loss of partner and we have a
medium ability to mitigate the threat. Right plot indicates that the loss of the partner
will have a low impact and that there is a high likelihood that we can mitigate the impact.

FIGURE 6.2
Plot of "probability versus ability to mitigate/capture" and of "impact versus ability to miti-
gate/capture." These plots combined with the "probability/impact" plot shown in Figure 6.1
are the recommended alternatives to the traditional Boston Square.

One more thing regarding how to deal with soft risks. As I mentioned in
the beginning of this section, a soft risk can be partly defined as one that
does not typically show up as a line item in the run-of-the-mill spreadsheet
model. I might use "fatalities" or "religious implications" as illustrating
issues that would bring a level of discomfort to most people if they were
tasked with filling in yearly costs in spreadsheet cells. Here's my advice on
that.

I have already suggested that truly sensitive soft risks be separately addressed—not ignored—and a unique assessment delivered to decision makers who will conceptually integrate the separate-assessment information. For those items that are not included in such unique assessments, my stern advice is to "suck it up" and enter line items and values for all remaining soft risks.

Typically, those lacking in intestinal fortitude will "hide" soft risks in already-existing line items. For example, a person might include the quantitative impact of "religious implications" in the costs associated with materials or other more mundane line items. This is a mistake.

If you are to include in your RR threat-mitigation actions or steps to capture opportunities, then those actions have to be associated with specific risks. Those risks need to be included in any model uniquely if their impact on success and value are to be determined. So, if a soft risk does not warrant being part of the unique risk analysis—and the vast majority of soft risks should not—then "be a man" and create a line item in the model that specifically deals with that issue. It might make some people squirm to set yearly values to these issues, but if you are the RPP, that's your job!

Have the Conversation about Risk Acceptance

In the previous section, I have partly—and sort of sideways—addressed the issue of risk acceptance. I did this when I mentioned the low-probability/high-impact risks such as the commercial airplane falling on the factory. I'll here expand upon the subject in the context of what we would like to have in a perfect world.

I'm a dad. With my wife—and it was mainly my wife—we raised two girls and a boy. I'm not without fault, but I think I was a pretty good dad—or was I?

I know that wood burns. I know that every day there are multiple house fires and some houses burn to the ground. Fatalities are often the result. Knowing this, would a good dad put his children—every day—in a structure made of wood with fires in the furnace and water tank, knowing fully well that burning to the ground and loss of the children was a possibility? What kind of guy is that??!!

I could have built a house completely out of titanium alloys that ignite only at exceedingly high temperatures. I could have done that, but I didn't. Why not? Truth be told, it was economics. You could make the case that I (and you) did not think my children's lives were worth the cost. Or was the matter more complicated than simple economics?

A contributing factor is the probability of your particular house catching fire and the loss of life. Although you know that it happens regularly, you decide to side with statistics (odds, if you will) and not spend the money on the titanium alloy house because you convince yourself that the prospect of a fire is exceedingly unlikely. Also, we install smoke alarms, make exit plans, and so on to help alleviate our angst.

And so it is with risk acceptance on a project. Why don't we build huge metal domes over our factories to protect against the falling commercial airliner or a meteor or space junk? If it was absolutely cost free to do so, we might actually build such domes. However, our decision not to construct such a barrier is based both on economics and probability.

Any risk analysis records risks that have some reasonable likelihood of happening and that would not bankrupt the company to mitigate or capture. Where the line is between risks that will be addressed in the typical risk analysis and those that will simply be accepted (nothing done to mitigate or capture them) is as individual as the project. However, a conversation and resulting documentation are steps that are absolutely necessary.

Such a conversation should be between the RPP, the PM, and the top-level decision makers. Only those to be ultimately held responsible for tragedies can inform where is that line between risks that will be assessed, monetized, and managed and those that will simply be accepted.

Having such a conversation requires some "out-of-the-box" thinking. I doubt that anyone would seriously discuss the commercial-jet-falling scenario unless the facility under consideration was near an airport. However, contributors to the conversation should certainly attempt to identify as many relevant "Black Swans" as is practical. Also, there should be a conversation about monetary limits. That is, how big is too big to be considered in the typical risk analysis? Similarly, how small a probability is too small a probability to be included in the general risk assessment?

This is an important conversation to have for at least two reasons: First, it makes decision makers acutely aware of some of the high-impact threats that might befall them and of some of the too-hard-to-capture opportunities that they will forego. Second, it allows risk modelers to ignore such outlier risks and make the assessment model as clean and comprehensive as is practical. Documentation of accepted risks and risk levels might save the reputation of the RPP should one of those risks occur.

Determine What Has Already Been Done and Establish Rapport

It would be a strategy lacking in foresight to approach a project team and upset the apple cart completely, so to speak. In the section of this chapter entitled "Try to Be Unfashionably Early" I submit that an RPP should attempt to be an integral part of a project team as early in the project lifecycle as possible, but I also admit that an RPP's participation in the very initial stages of a project is unlikely. Regardless of the stage of entry, advice in the following section holds.

Find Out What Has Already Been Done

Even speculating that you can override any or all of already-existing risk processes and models with a new properly risk-based implementation is silly. When an RPP approaches a project team—any project team—he/she will find that there are at least risk-like practices already in place in some of the contributing disciplines, and, most likely an initial commercial model (spreadsheet) has been constructed.

Sometimes uncovering all of the in-place practices can be disheartening. Depression can stem from the fact that some of the "risk-based" practices are just plain wrong. Another source of the blues can emanate from realization of what it is going to take to integrate the various existing methodologies or to influence them such that they can be used as critical elements of a comprehensive risk analysis. When I teach classes, at this point, I regularly relate my "backstroke" story.

If you have never performed the backstroke when swimming and I am attempting to get you to do it the way I do it, then it is likely that you will take some time to learn the stoke, but in the end, you will mimic my method. Contrast this with attempting to teach someone who has been "backstroking" for years but does not do it the way I would like them to. Attempting to change that person's method likely is going to be a more onerous task compared to the teaching effort required to influence the uninitiated.

It is a rare instance that you can walk into a truly "virgin" situation with regard to risk practices. So, one of the initial steps to be taken is to survey the landscape and find out what is already going on. Document every detail, compile the information, and most importantly, sit on it and think about it for a while before opening your mouth.

If a spreadsheet model has already been constructed, sit down with the author(s) and determine how it works. Document what potential risk items they already have incorporated in the model and get a feel for whether or not they would be receptive to adding new parameters—especially those that represent soft risks. Listen to their feedback, and, again, think about it for a while.

This "for a while" stuff is important. First, it gives you a chance to get things in perspective. Second, it gives the parties you interviewed a chance to mellow to the idea of change. Their first reaction is almost always going to be more harsh than will be their position a few days later.

Go Back to the Same People with a Plan

In the "for a while" period, construct the best compromise plan you can conceive. For example, if one group is using traffic light analysis (red, yellow, and green lights to represent high, medium, and low risk, respectively), you might devise a proposal in which they still evaluate risks in the traffic light manner. You might suggest an addendum to their method in which you convert their traffic light color to quantitative data that can be used in the comprehensive risk-monetization process.

For example, you can devise a proposal in which for each traffic light, you gather information on the following:

1. What is the likelihood that the risk might occur (can collect this as a range of minimum, peak, and maximum likelihood)?
2. If the risk does happen, for how long will it delay us (can collect this as a range of $ per time unit in the same minimum, peak, and maximum manner)?
3. How many $ per time unit will it cost us (can collect this as a range of minimum, peak, and maximum $ per unit time)?

Now, the traffic light can be converted into a quantitative assessment by the simple equation:

$$\text{Range of cost} = (\text{Range of probability}) \times (\text{Range of delay}) \times (\text{Range of \$ per time unit}).$$

Such a proposal is a reasonable compromise because it allows the discipline in question to continue to utilize their traffic light process while affording you the quantitative data required. Other conversions from already-existing "risk" systems to quantitative analysis likely will be required.

After you have conjured up hopefully amenable methods to either replace already-existing methods (sometimes you can't save them) or to shoehorn the existing methods into the comprehensive process, you should revisit the same individuals. They should be presented with the overall plan, should be appraised regarding what changes if any will be required to incorporate their methodology, and should be impressed with the critical role represented by their contribution to the overall process. Pay attention to their feedback and attempt—to the extent possible—to accommodate their suggestions.

Be Viewed as a Trustworthy Advocate

One of the cardinal rules of building trust is to refrain from promising things you can't deliver. Not much is seen to be closer to an outright lie than breaking a promise or reneging on an implication. It is often the case that RPPs get a bit carried away with the story about "how wonderful it is going to be" and that is a mistake.

Deliver bad news yourself. If a particular group currently is utilizing a risk practice that just is not going to fit into the comprehensive scheme, it is bad practice to have them hear it through the grapevine, so to speak. Bad news should be delivered in a timely and cordial manner and not by a proxy.

Be honest about the benefits of adhering to the proposed risk process. It probably will translate to more work on their part. The benefits might not be immediately evident. In fact, some groups will actually have to "fall on the sword" a bit for the overall good. If a group has to sacrifice (take on work, for example, that won't necessarily benefit them directly), make it clear why the sacrifice has to be made and attempt to put them in touch with those who will benefit. In most instances, individuals and groups will comply if they are confronted with the success story and meet the people who will benefit from their extra work.

Don't get into the "because the CEO says we have to do it" mode. The carrot of logic and compassion always yields better results than the stick of "because the boss says so." Using the stick might result in compliance, but it will be the "check-the-box" type of adherence to the process. If people actually want to participate, they will adhere to the spirit of the law rather than just to the letter.

Establish a High-Level Story about Why This Is Being Done

If you were to just show up at someone's door and suggest they change the way they currently do things, your prospect for cooperation would be poor. Change, even for the best of reasons, can be difficult to do and can be more difficult to sell. Advice in the following sections might help.

Appeal to Their Sense of Logic

As the old adage goes, "simple is best." So, it is with examples of why risk analysis generally is a boon to the success and value of projects. Unless you have an acute case of the "I just don't want to know" syndrome, it is tough to imagine that it is better to be ignorant than to "see" problems coming. One of the salient features of any credible risk analysis is that of attempting to look down the road, so to speak, of the project lifecycle and to predict what problems might arise and/or what opportunities might appear.

When armed with the knowledge of what conceivably might happen, it doesn't take a genius to tumble to the conclusion that taking steps now to avert the threats and to capture the opportunities—as best as is practical—will yield a more trouble-free and successful project. Of course, the counter argument is that the risk process itself is a work-creator and a cost. True enough. However, if just one major problem is avoided or one significant opportunity is captured, the penalties of enacting the risk process are more than offset.

Well, that's all very nice, but I have actually had people relate to me that all of that could be accomplished without their participation. The counter to this argument is integration. In the examples from the, "Dashboards and Traffic Lights" and "It's about Time" subsections of "Surprise! Integrated Impact Is Almost Always Alarming" in Chapter 1, I illustrate clearly that it is only through the integration of risks emanating from all disciplines that the best prediction of success can be made. This leads me to the next section.

Appeal to Their Egos

Chances of failure are like links in a chain—if one of them breaks, that's it. Consider the example of wishing to take a trip in a car. For that to happen, the engine, exhaust, cooling, electrical, breaking, transmission, systems, and others all have to work. For any given trip, there is some probability that the engine will fail. For the same trip, there also is some probability that the exhaust system will fail and an independent probability that the cooling system will fail, and so on. It matters not if the engine is great and the transmission is great and the cooling system is great if the electrical system is shot. The point is, failure of one component causes failure of the entire enterprise.

In Chapter 1, the example given in the "Dashboards and Traffic Lights" section is especially representative of this point. In that example, all traffic lights are green in color, indicating a less than 10% chance that any given discipline will cause project failure. However, there are tens of disciplines. Statistics would dictate that at least one or more of those disciplines will cause failure (if we had 10 disciplines each with a 10% chance of causing failure, then statistics would dictate that, on the average, one of those disciplines would fail).

Nothing could demonstrate better that you can't generate a reasonable probability of overall success unless each discipline contributes. So, success of the entire risk process depends on them (fill in the name of the discipline you are trying to convince) accurately representing their situation. In the "It's about Time" section of Chapter 1, I illustrate that it is only through the integration of all independent and dependent time projections can we get a realistic estimate of, in the case of the example, starting time.

Each of these examples and many more that I simply don't have room in this text to convey, illustrate the principle that a credible risk analysis can only be performed when all major contributing parties are accurately represented. In other words, when you talk to the skeptical, you can honestly say: "We can't be successful without you!"

Another logic point is the interdependency of disciplines. Let's say you are in discussions with the Security folks. Typically, in the past, a Security risk analysis consisted of a textual report documenting likely avenues of breach and internal and external physical threats. Such a report was delivered to management and that was that.

Building on the "Dashboards and Traffic Lights" example in Chapter 1, the case can be made that examining the report of any discipline in isolation does not give an accurate picture of the chance of success. Disciplines impact one another. For example, it is relatively simple to convey the logic that the Security risk assessment will impact other disciplines such as logistics and environmental affairs. Unless Security risks are clearly defined in Reason/ Event/Impact (REI) format and properly quantified, it can be difficult for representatives in the Logistics Department to make logical decisions regarding how to safely transport materials to and from the project site. Similarly, experts in the Environmental Affairs Department need security information to make logical determinations regarding the placement and protection of waste pits. So, the argument that "It all depends on you!" is not only flattering, but mostly accurate.

Be Ready with Stories of Success

Sometimes, I'm just an idiot. When I first started messing around with this risk stuff—back when the world was young—I simply ran "pillar to post" in an attempt to help one group after another. Most of my efforts were successful—at least on some level. I'd be embarrassed to tell you how long it took me to realize that I should be collecting "post mortem" data on those projects. That is, I should have from the beginning been documenting the types and levels of success, including testimonials from satisfied clients.

Don't be like me (generally good advice)! Make sure you document your successes and failures with regard to implementation and execution of risk processes. As I have said repeatedly in this book, nothing sells like success. You should always have at the ready at least a few stories to which the listener can relate.

If you don't yet have a war chest of success scenarios, then search the literature to find documented victories in other businesses like yours, or, in businesses not at all similar. Or, you can contact some old grizzled guy like me and I can bend your ear for hours. In any event, collect success stories and commit them to memory. Most importantly, revisit groups with whom you have had a risk-implementation/execution experience and collect quotes from them.

Design a Risk-Monetization Process That Fits with the Existing Culture and Workflows

In preceding sections ("Find Out What Has Already Been Done" and "Go Back to the Same People with a Plan"), I have suggested practices that can be used to modify—or, sometimes completely override if necessary—the already-existing risk-based practices of any given discipline. In the physical world, an analogy might be the building very different individual train cars so that they might be able to be linked. Assuming success in this endeavor, it is now the place in the process to ponder in what logical order the cars will be connected to create the train.

If you are an employee of any discipline, it is upsetting enough to either have a brand-new risk process foisted upon you, or, to have your tried and true risk process modified in any significant way. Given that only through integration of risks from all disciplines do you achieve a most practical perception of project success and value, the task of the RPP is to devise a logical "road map" that will connect the disparate processes. This effort represents a chance to even more cause consternation in the ranks. Therefore, it is best practice to design a risk integration process that "looks" as much like the existing interdepartment workflows as is practical.

Understand the Workflows

In the "Integrated into the Way We Work," section of Chapter 1, I offer two brief paragraphs that simply serve as a warning that consideration of workflows is an issue. In this section, I will elaborate on how such consideration would work in the "perfect world" scenario and will give examples.

A workflow is a series of steps that are typically followed to achieve a specific goal. For example, it might be that Company X is in the business of constructing manufacturing facilities in foreign countries. The typical workflow for establishing a facility might be the following:

- Negotiators make initial contacts with foreign entities and decide whether or not a reasonable deal might be struck.
- If a reasonable arrangement might be agreed upon, the Legal team evaluates the legalities of all aspects of doing business in the country.
- If legal matters are not an issue, the Security team evaluates the practicality of establishing and protecting a facility.
- If a reasonable deal can be struck and there are manageable legal and security concerns, then the Health and Safety department looks into medical, health safety, and other matters.
- If the proposed project passes evaluation by all the aforementioned entities, then the Design team generates alternatives for construction, supply chain, logistics, etc.
- Commercial analysts will evaluate each design and recommend that which they deem best.
- When a specific design has been selected my management, the Construction and Logistics department collaborate on building the facility.
- Following construction, the Operations department assumes control of the facility.

Now, nobody has to rub my nose in the fact that the steps outlined above represent a grossly oversimplified version of an actual workflow for such a project. The list is meant to be illustrative. The point is that no matter how large or small the ambition, there likely exists a predetermined workflow—or set of steps—through which the project will pass.

It has been my experience that attempting to significantly change the workflow is tantamount to standing on the tracks and attempting to get the oncoming freight train to take a different route. You can stand there if you like, but the result usually is not pretty.

"Understand the Workflows" is the title of this subsection, and that is stellar advice. Regardless of how much you, as the RPP, are able to influence the specific risk process utilized internally by any of the participating disciplines (negotiators, attorneys, security analysts, health and safety staff, designers, commercial analysts, logistics experts, construction, and operations personnel), creating those more practical and utilitarian risk methods will mainly be for naught if you then recommend that those methods be "strung together" in a workflow that is unnatural to the participating entities. Don't stand in front of the train!

It will be nearly inevitable that you will see efficiencies that might be gained if the workflow could be changed. Unless that change is particularly minor in execution and effect, my recommendation is to create a risk-assessment/management/monetization process that abides by the existing linkage of disciplines. For example, it might be inconvenient to consider the commercial evaluation of the project prior to determining just how the project will be constructed and how critical logistical aspects will be handled. In such a case, the best advice is to simply recommend that the commercial analysis—with its identified threats and opportunities—be revisited after the construction and logistics plans are known. This represents a slight twist in the established workflow, but likely is a step they take anyway even if it is not a "box" on the workflow chart.

Understand the Incentive System

As I have mentioned numerous times in preceding sections, this chapter is about what might be done in a "perfect world" scenario. That is, how would you have it if you could dictate every step? No other influence is more detrimental to the "perfect world" plan than the existing incentive system.

In the "Organizations That See the Benefit of Doing This—The Incentive System" subsection of Chapter 1, I state: "The incentive (reward) system typically works against risk-process implementation." I also remark that: "The bottom line is that incentive systems in most organizations are not set up to favor the risk-process approach."

In that subsection of Chapter 1, I give an example of the lack of incentive for groups who handle the project early in the lifecycle—when the initial RIW is convened—to address threat-mitigation or opportunity-capture action items for threats or opportunities that might or might not materialize "down the road" on someone else's watch. Although this illustration of the juxtaposition of the incentive culture and the risk system is true enough, it represents just the tip of the proverbial iceberg.

In the "perfect world" scenario, project personnel would have incentive to adhere to the letter and spirit of the risk-process tenets. Sadly, this almost never is the case and the "strange bedfellow" relationship between the risk and incentive processes goes far beyond the example outlined in Chapter 1.

Businesses are bottom-line focused. They are in business to make money. One aspect of both making and saving money is efficiency. That is, efficiency in executing a task can directly or indirectly influence the profitability of an endeavor. Incentive systems typically are constructed such that increased profitability is rewarded. This means that personnel who find better ways to do things in the accepted workflow (see previous subsection) will be rewarded. Then, along comes the risk initiative.

As I stated in a previous section, it is a rare instance if the risk process is considered when initially setting up the workflow for any process or project. Risk-related methods—especially those like the monetization process—nearly always come along after the workflow is well established. So, regardless of the possibly pitiful success rate of any arm of the company with regard to executing successful projects, that segment will strive to avoid connection with a risk process that is perceived to require additional work— which it does. I state repeatedly that "We often get rewarded for successfully launching a project, but not necessarily for launching a successful project."

In the "Be Ready with Stories of Success" section of this chapter, I relate how important it is to have at the ready examples of how application of a risk process enhanced project success and value. This is all very nice, but such yarns are not likely to precipitate changes in the behavior of personnel if the steps they might have to take will be detrimental to their personal reward.

For example, it might be best for the project if a report that includes risk documentation, be delayed in delivery so that a commercial risk assessment can be completed and incorporated into the report. Such delay might not be typical. If the person responsible for report delivery is rewarded for getting the report to recipients "on time," it is unlikely, just because the quality of the report would be enhanced that the person will delay delivery. "On time" is rewarded and "late" is punished. If you were that person, would you be willing to "take one for the team" and wait to include the commercial analysis?

I am here neither suggesting that an RPP can—or should even try to— change the way all things are done so that risk processes can be accommodated nor am I advocating that an acceptable case can be made to personnel to act in direct discord with the existing incentive system. However, I am stating that any RPP has little of no hope of devising a risk process that will gain acceptance unless that RPP has a relatively comprehensive understanding of the existing incentive system. Any process promoted should recognize and "blend" with pervasive reward practices. This, I know, can be a real challenge.

This issue is an example of the type of dilemmas that should be initially discussed with the PM. It should be made clear to the PM that some of the recommended methods likely will "fly in the face" of the current incentive system and will undoubtedly be viewed as extra work. It is the job of the PM, to the extent that he/she can, to convey to project team personnel that adherence to the risk practices will be rewarded and that the "extra" work represented by the risk process should no longer be viewed as an "add on" to the existing workflow but should be now be considered as an essential part of a new workflow. The PM has to not only say this, but has to mean it. Early examples of reward for adherence to the risk process always aids in acceptance.

Establish Common Terms and Definitions

In a movie I once saw, one of the characters repeatedly used the term "inconceivable" and another character eventually remarked to the first: "I do not think that means what you think that means." After lack of effective management support, I would rank communication problems as the second greatest cause for risk-process-implementation confusion, consternation, and failure.

I have addressed various aspects of the communication issue previously in this book. In the "Communication" subsubsection of Chapter 1, I remark on colloquial verbiage and the differing views held by disparate disciplines regarding the definition of terms such as "risk," "uncertainty," and others. In the "Poor Risk Communication" section of Chapter 5, I delineate the importance of establishing a chain of communication and coming to consensus regarding just who will communicate what to whom.

In those sections, I primarily indicate what are just a few of the foibles related to communication. Given that this chapter is focused on how things might be handled in the perfect world, I will put forward at least one suggestion that goes a long way toward being a remedy for myriad communication issues.

Being of German descent, I took German language classes in high school and college. In the usual ignorant/arrogant view of youth, I thought I had a pretty good grasp of the language—that is, until in my twenties, I made my first trip to Germany. I could read signs OK and could make myself haltingly understood by most German speakers. However, I received a lesson in life when I realized that I could not understand one word they said to me. The rapidity with which native speakers (even those of us who profess to be able to speak English) rattle-off their language was something for which I simply was not prepared. More than that, though, my inability to understand was due to the fact that the German that is taught in school (in the United States) is not the German language people actually speak.

Turns out, there are many dialects. Newspapers, magazines, and the like are printed in "Standard German"—a "dialect" that is not necessarily spoken, but that is universally understood. Later in life, I tumbled to the realization that the "Standard German" precept could be applied to matters of risk communication.

It is folly to expect that you will have any success at all in attempting to convert the denizens of any discipline to change the way they utilize terminology related to that area of expertise. Good luck, for example, with persuading personnel to embrace a new definition for an already accepted and ubiquitous term. All is not lost, however, if you take a lesson from the German language.

You might not be able to alter colloquial definitions of terms such as "risk," "uncertainty," and others, but you can—and I have—successfully establish a lexicon of terms and definitions that will be used in written interdiscipline and "up-the-chain" communications. In this model, you don't even attempt to change the "dialect" of any particular group, but you come to consensus regarding what will be meant by terms when communications are intended for outside-of-the-group consumption.

This works. It appeals to various groups because they can carry on day-to-day operations sans having to alter their speech. It appeals to management because managers do not have to learn what "Group A" means, for example, when they communicate the term "mitigation" and how that differs from what "Group B" means by use of the same term. Coming to consensus regarding critical terms and definitions is absolutely essential if you are to be successful in implementing a risk process. The "Standard German" model is a relevant remedy.

Establish That There Will Be Only "One" Risk Register

This is one of those glaring examples of "talk is cheap." It's frivolously simple for me to "say" here that only one RR should be allowed to exist on any project. Causing that to happen, however, can be anything but simple. So, what are the hurdles?

In the "One Lump or Two (or More?)," "Who Has Access?," and "Who's the Boss?" subsections of Chapter 5, I address just a few of the conundrums that plague the establishment of a single RR in the contemporary and typical situation. In this section, I expand on those initial precepts to illustrate for what you should strive in the "perfect world" scenario with regard to RRs.

A perfect world would be free of litigation, industrial espionage, jealousy, and other sources of strife that plague the planet. In the perfect world scenario, the legal department would not be concerned that legal strategies or assessments of partners would be shared outside a very small circle of company attorneys and management. However, the escape of legal views and analyses, that can be integral parts of risk documentation, absolutely is the sort of stuff that keep company lawyers awake at night. Similarly, commercial analysts constantly fret regarding the "liberation" of competition-beating strategies, health and safety personnel need to guard against publication of the financial aspects of life-and-death decisions, and so on.

When a risk-identification workshop is held, information about the aforementioned and many more sensitive issues are directly or indirectly captured in the Reason/Event/Impact description of threats and opportunities. Therefore, the RR can be viewed as a real powder keg with the potential to disclose matters that could significantly damage the company.

It is not such an onerous task to get, say, the legal staff to create an RR in which they record sensitive information. The trick, of course, is to convince the legal group to relinquish ownership and control of that register in favor of a single RR that is controlled by someone outside the legal department—namely, by the RPP. The RPP will be, in the perfect world scenario, the gatekeeper for information and data flow in and out of the RR. It will be the RPP, for example, who adds to the register new risks as they materialize and who passes judgment on whether or not threat-mitigation or opportunity-capture actions have been effectively carried out so that a risk can be removed from the register.

My message here is to the RPP. Because the risk process is, at its best, holistic, it is dependent upon the contribution of all pertinent disciplines, is backed by project management, and is central to the success of the project. Therefore, it is very easy for an RPP to get wrapped up in the importance of the risk process. An RPP might begin to believe that everyone should "see the light" and simply comply. If that's what you believe, then I've got some swamp and I'd like to discuss with you.

Try to put yourself in the shoes of the other guy. You are the generator and keeper of highly sensitive information and data that could have disastrous consequences for the company. Moreover, you are responsible for the safeguarding of such information. Why would you even think about relinquishing control of that information to another person such as the RPP—a person you likely don't know well? Why would you view it as a good thing that our portfolio of sensitive facts should be allowed to be "mixed" with documentation from other seemingly disparate disciplines? I'm here to tell you, you wouldn't.

This is one of the main reasons that risk processes die before they get off the ground, or, why they can be spectacularly unsuccessful. As the RPP, it is absolutely legitimate to be sensitive to the concerns of others regarding the security of information.

In the "Holistic Set of Integrated Risks" section of Chapter 1, in a broad-spectrum way in the "Health & Safety" section of Chapter 4, in some detail in the "Ignoring Soft Risks" section of Chapter 5, and in the "Deal with Them" and "Politics" subsections of this chapter, I delineated processes by which risks associated with exceedingly sensitive arenas, such as fatalities, can be separately assessed and then can, by management, conceptually integrated with results from the main holistic and comprehensive risk analysis. If each discipline is allowed to make and win the argument that their information is of such a sensitive nature that it deserves such separate treatment, then the goals of any risk-assessment/management/monetization process cannot be met.

If the RR with which the RPP is left is one that does not reflect the critical and, yes, sensitive elements of the project, then that RPP is relegated to lord over an anemic set of threats and opportunities that will have little material impact on the project or on the company. That is, the risk process will be assessing, managing, and monetizing a bunch of stuff that doesn't much matter. This can't be allowed to transpire.

One of the elements of the aforementioned discourses on the handling of sensitive risks is that of deciding—"up front" in the risk process—just which risks will be allowed to be given special and separate treatment. I make an example of risks associated with fatalities. The cohort of risks that are granted dispensation has to be very small. All other risks—regardless of the arguments presented regarding their unique and sensitive nature—have to be included in the general RR. This, once again, is a situation in which backing of top-of-the-house management is crucial.

In the "One, Two, Three—Go!" section of Chapter 5, I address the issue of wasting company time and resources in the "fixing" of risks (mitigating threats or capturing opportunities) that will have little to no material impact on the project. For example, the "worst" threat from department A might have orders of impact less potential impact than the "least" threat from department B and, therefore, possibly none of the threats from department A warrant attention. If separate RRs are allowed to be kept by each or by most disciplines, then you can be assured—because all risks can't be practically ranked—that such inefficiencies will be rampant. The keeping of one RR with one "point of entry"—the RPP—is essential.

It ought to be obvious that in making and winning the argument for a single RR, the RPP's neck is sticking way out. If, as the RPP, you are able to convince the various disciplines that the information with which they have entrusted you is safe, then you had better be sure that your assurance of safety is valid. This assurance goes well beyond issues of your personal integrity. Physical security, IT security, and trust in coworkers who will aid you in management of the RR are just a few of the areas over which you might not have much control. Be sure that all of the elements that could be the source of an unauthorized breach of the RR are, at least, up to the generally accepted company standards.

This leads me to my last point in this section. With the advent of modern information-management techniques, it is possible to, in "cyber space," cause separate entities to appear as one. That is, it is now possible to create one risk register image that actually is the electronic confluence of disparate RRs that are stored in physically unique locations. I'm an old guy and I freely admit that I can't begin to understand the nuances of the modern data-management world. However, I also don't have any qualms concerning admission that I would, personally, put my limited trust in the architecture that offers the fewest avenues for potential breach. That's all I have to say about that.

Prepare for Risk-Identification Workshop

Quite a bit of this book is focused on the risk-identification workshop (RIW). There are several good reasons for that. First, the process of preparing-for and actually convening of the RIW will define the relationships between the principal players—mainly the PM and the RPP. In addition, the RIW is the forum—hopefully convened temporally near the inception of the project—in which most major threats and opportunities and all attendant information will be captured. If this step is not executed with authority and skill, there is not much hope that the remainder of the risk process will have a happy result. So, in an effort to guide you toward a satisfactory result, I offer the following words of wisdom.

Make Sure That the RPP and PM Agree on "What Is the Question?"

In the "For Absolute and Relative Purposes" subsection of Chapter 1, I describe that although each member of a project team would swear that they are all working in service of the same project, their individual views of the project—and therefore, the main goals of the effort—are divergent. In that same section, I offer a practical approach to causing confluence of views and, in turn, consensus on just what question(s) needs to be addressed by a risk assessment. The "Be Clear as to 'What Is the Question?'" section of Chapter 3 address how the perception of the project goals of any given team member can be very different from that of the PM, how temporal aspects impact the answer to "What Is the Question?," and how the different perceptions of the project held by representatives of various disciplines can impact one another.

It might be true that you have gone through the exercise I describe in Chapter 1 to come to consensus regarding just what problem the project is supposed to address and precisely what is to be expected from the risk analysis. However, that seeming confluence of viewpoints must be carried through to the expression by the PM of the project goals at the opening of the RIW.

I have been around a long time and have dealt with many a PM. I can tell you from experience that it is typical to believe that you have come to consensus with the PM and other members of the team regarding the goal of the project and of the risk analysis only to find that when the PM expresses his/her view to the team at the opening of the RIW, that view is the same one they harbored prior to coming to "consensus."

Allow me now to backtrack just a bit. At the RIW, following introductions of the attendees, the PM—or a designated proxy—should stand before the entire gathering and state clearly just what constitutes the project, exactly what are the goals of the project, and precisely what role the risk process will play. I mention here that sometimes a proxy—that is, a designated speaker—has to deliver this message for the PM. If you have been associated with projects for any time at all, you will be well aware of the fact that some people can nearly flawlessly execute the day-to-day responsibilities associated with project management, but those same people might possess no facility for public speaking. If this is true of the PM with which you are working, it is the job of the RPP to, kindly, make the PM aware of the fact that it is critical that the opening statements are delivered succinctly and that they (the PM) are not the person for that job.

Regardless of whether the PM or a substitute delivers the opening salvo at the RIW, it is exceedingly common for that person to revert back to their personal view of just what is the project. In addition, the PM often will attempt to impart to the gathering what are his/her major concerns and goals. So, even if you have to prepare a text that has to be read (speeches that are read are almost always deadly from the point of view of the audience), then that is what the RPP should insist upon.

For example, it is common that the PM is mainly concerned with personnel issues, budget, and schedule. I know that this is to grossly oversimplify the scope of concerns for the PM, but please allow me this distilled analogy for the sake of example. The negotiator in the audience, however, is concerned mainly about his/her efforts to hammer out a high-level deal with the host government and all subordinate governmental agencies. The commercial analyst is primarily concerned with whether or not there is a practical market for the potential product and with the positioning of competitors. Whether or not a potential production facility can be safely built and defended is the primary focus of the security analysts in the room. And so it goes.

A PM-point-of-view message delivered to that diverse crowd will ring hollow with nearly every member. Any expression of the goals of the project—the answer to "What Is the Question?"—absolutely has to incorporate elements of the concerns from each major discipline represented at the RIW. That is, the scope of the project presented needs to recognize the timely contribution of all major area of expertise and should illustrate how these disciplines are linked in the project life to accomplish the main objective of the project.

Just for example, a PM (or skilled-speaker proxy) might start the meeting with a statement that begins—in part—with

> Good morning. You are sitting in this room because you represent a critical element of the New Factory project. Each of us will have a crucial role to play in the successful execution of this multi-million-dollar effort.

> It is my intent, now, to express to you the scope of the project and the major objectives that will define success.
>
> Our negotiating and legal teams, lead by Bill B. and Sally S. respectively, have had initial talks with the host government in an attempt to assess the feasibility of establishing a manufacturing facility in Foreignland. Simultaneously, our commercial team headed by Jane J. is evaluating the economic viability of producing our product in Foreignland and what market exists. Bob B. of the logistics team is currently assessing the practicality of getting raw-materials to, and finished-product from the factory site .... .

With such a presentation, each member of the project team is made aware of the contributions of other entities and their particular place in the scheme of things. I know you might think it might be normal, for example, for a member of the legal staff to be aware of the relative contributions of the logistical element of the project, but, believe me, it is not normal. Such a speech creates a sense of community and of codependence. A stellar presentation will emphasize the fact that each discipline gets information from other disciplines and also "feeds" areas of expertise that are temporally subsequent in the project lifecycle.

A well-constructed opening oration might conclude with a statement of the high-level goals of the project such as

> Although each of you has a specific set of objectives to be met for successful execution of your part of the project, let me now present to you just a few of the high-level goals. Suppliers for our product are crowding the market. Company Y—our major competitor—in fact is evaluating construction of a new manufacturing facility that would, if we don't beat them to the punch, target the same customer base as we have targeted. Beating Company Y to the market, therefore, is a primary goal of this project. In addition, you are all aware of the financial and personnel constraints under which we currently labor. The budget for this project is both finite and well defined. Most efficient execution of the project depends on many things.
>
> One major element of timely and efficient execution is to foresee the threats and opportunities associated with the project and to mitigate those threats and capture those opportunities in a cost-effective and timely manner. This is the reason-for and focus-of our risk assessment/management/monetization effort.
>
> Another critical aspect to success of the project is the efficient interaction of the many disciplines represented in this room. The risk process is designed to facilitate such efficiency .... .

It is essential that this opening diatribe be delivered in an earnest manner and with a sense of urgency. Again, if the PM is not the sort of person who can deliver such an oratory, then get someone who can. These preparations should be done well in advance of the RIW.

Prepare for Risk Expression

In the "Risk Description—How to Express a Risk" subsection of Chapter 2, I illustrate the importance of utilizing the REI format to describe a risk. In the "Appeal to Their Egos" subsection of this chapter, I describe how the REI format facilitates realization among disciplines how any area of expertise might be linked to a particular risk. Also in the "Establish That There Will Be Only 'One' Risk Register" section of this chapter, I point out that due to the comprehensive description of risks necessitated by the REI format, the RR can be seen as a real powder keg—that is, a place where the sensitive nature of any risk is detailed.

Regardless of agreement regarding adoption and adaptation of the REI format for capturing risk descriptions, unless you employ a mechanism at the RIW, which forces utilization of the REI format, attendees will revert to less rigorous means to document risks. Therefore, some preparation is in order.

Just to refresh your memory regarding why the REI format is important, what we are trying to avoid are curt and nearly useless expression of a risk such as "logistical problems." Such a threat should be expressed in REI format thus: "If our negotiations with the host government are not successfully completed before October of this year (Reason), then because we will have missed the 'weather window,' we will have to delay shipment of materials until spring of next year (Event). Such a delay will add millions of dollars to the capital expense (CAPEX) budget and will cause a delay in schedule of at least 6 months (Impact)."

Opportunities are similarly handled. Without employing the REI format, an opportunity might be expressed as "lower costs" or by some other equally general statement. In REI format, such an opportunity might be stated as, "If we can successfully negotiate a contract with Vendor X for supplying steel (Reason), the significantly shorter shipping distances (Event) will drastically reduce our cost of raw materials (Impact)."

Such detailed and succinct expression of the risk also allows project teams to design threat-mitigation or opportunity-capture actions that address either the probability of the risk happening, the impact of the risk if it does happen, or both. It is absolutely essential to establish whether actions to be taken are in service of affecting the probability-of-happening or of impacting the consequence of the risk.

I have been witness to all manner of pre-RIW preparations to facilitate utilization of the REI format for capturing risk descriptions. On the low-tech end, it is typical to generate preprinted 3×5 cards with Reason, Event, and Impact vertically stacked on the face of the card. High-tech solutions include spreadsheets or other electronic mechanisms, which "force" the sequential addressing of each element of the REI format. Regardless of the mechanism chosen, you should absolutely not depend upon the good graces of the attendees to utilize the required format. Prepare your risk-capture mechanism (3×5 cards, spreadsheet, or whatever) well in advance of the RIW and require use of that mechanism.

Who Should Attend and How to Invite Them?

I have addressed some aspects of this subject in the "Incomplete Representation" section of Chapter 5 and in much detail in the "Establish the Powers Bequeathed to the RPP" subsection of this chapter. In those preceding sections of this book, I pretty much cover the salient aspects of this part of preparing for the RIW, so I won't dwell on this subject here. However, this facet of RIW preparation is important enough to warrant a few words in this section.

It should be agreed, well in advance of the RIW, just who will be "invited" (read "required") to attend. Such agreement should be between the PM and the RPP.

An RPP typically is "a new guy in town" or, at least, the RPP position is relatively poorly understood. Therefore, any invitation to a potentially multiday gathering such as the RIW is likely to be ignored or be tagged as low priority by most invitees. If the invitation is to be taken seriously, then such a solicitation has to be issued under the signature of both the PM and the RPP.

Invitations should not be issued by the RPP alone mainly because, as stated above, such requests will lack the required heft to entice participation by senior staff. The request to participate should not go out under the signature of the PM alone. Such invitations undermine the credibility of the RPP. It must appear to the staff that the PM and RPP are of one mind on this matter and that ignoring the solicitation will not be tolerated. No specific threat or negative consequence should be referenced, but the implications of noncompliance should be clear.

Determine Main Risk Categories and Who Will Address Them

Every project is different (profound, huh?). Therefore, there is no way I can here encapsulate the major sources of risk for a "generic" project. All I can do is give an example and leave it to the reader to extrapolate that example to the project at hand.

As stated many times previously in this book, a risk process needs to be one that is holistic. That is, it is critical that all pertinent disciplines and areas of expertise participate. So, for a typical project we might expect input from the legal, logistics, political, negotiations, commercial, financial, health and safety, security, design, research and development, construction, marketing, audit, and many other areas.

At the RIW, it is necessary to categorize risks at a higher level. That is, you should not attempt to capture specifically "legal" risks in their own category separate from "logistical" risks, and so on. It is far more efficient to establish high-level categories that pertain to the project and to populate those categories with the risks expressed for specific disciplines.

For example, for a project that is aimed at designing, building, and launching a commercial-grade software application. A typical set of categories might be the following:

- Location and Business Type
- Commercial/Competition
- Design/Engineering
- Information Technology Innovation
- Installation/Maintenance

Again, I am in no way stating or inferring here that this list of high-level categories be adopted by projects, or, even serve as a basis for modification. I am simply trying to impart to the reader that a relatively small number of broad risk categories should be established prior to convening the RIW. It will be decided—on the fly, so to speak—at the RIW to which high-level category each specific risk belongs.

There are several reasons for taking this approach. One is that, just for example, a legal risk might actually impact commercial aspects of the project and have nothing to do with the Location and Business Type category or others. It is important for the summarization of RIW results, therefore, to assign each discipline-specific risk to one of the preestablished high-level categories. If this is not done, reporting of results, the structure of the RR, and the tracking of actions becomes impractical.

Another reason for establishing high-level categories is to bring to the RIW participants the integrated nature of the project. In any high-level category, the lawyer, just for example, will see and realize that legal risks are codependent and linked to risks from many other disciplines, the risks from which have been assigned to the same high-level category. Anything you can do to impress upon participants the holistic and integrated nature of the project is of great advantage.

As will be described just a bit later in this chapter, the RIW will be organized around the major high-level categories. Such organization necessitates the segmenting of the RIW population into as many cadres as there are high-level categories. Prior to the RIW, it should be agreed between the PM and the RPP which invited participants will participate in the category areas.

For example, when considering who will participate in the IT Technology Innovation category, it is good to have at least one participant who is intimately familiar with the technical aspects of the project. However, the population of participants for the Information Technology Innovation category should not be represented by those with strong technical backgrounds. Commercial analysts, lawyers, and the like should constitute most of the Information-Technology-Innovation-category participants.

This might at first seem counter-intuitive. However, such a mix of participants in any of the high-level categories ensures several things. One is that most of the participants will be out of their comfort zones. This facilitates the discussion of issues that would ordinarily be taken for granted by a group of experts in any given field. This is an important point. I have served as the risk expert in countless workshops and general discussions with discipline experts. When a group of experts on a given subject meet, much is taken for granted and is left unspoken.

Just for example, when meeting with a group of construction engineers involved in a construction project, it can be the case that their conversation begins with the construction elements of the project—as you might guess. Being the ignorant one in the room, I have often piped up and asked questions about the timely delivery of raw materials or about the quality or status (union, nonunion, etc.) of the workforce or about the ability to move workforce personnel around on the job site or to-and-from the site, and so on. Given that I have no pride to protect regarding the subject matter, I have found that questions that are fundamental, that need of being addressed, and that would go otherwise unspoken are only considered if someone who is not a discipline expert is part of the mix.

Another positive aspect of such a mix of disciplines in an RIW category-discussion group is that it gives each participant a new appreciation for the high-level-category area. Still another advantage is that it causes people of different disciplines to meet and interact with one another. When is the last time, for example, you witnessed a mechanical engineer and lawyer talking (in civil tones) to one another? Any amount of cross-pollination is to the good. More on the technical reasons for having high-level categories will be addressed later in this chapter, but in a perfect world—which is what this chapter is all about—the PM and RPP would have prior to the RIW worked out the high-level categories and which RIW members will address each category.

Prepare Forms

I know this sounds mind-numbingly mundane, but before the RIW preparation of the forms to be utilized is an essential element of success. Disaster almost always ensues when either the format of forms to be used are designed and built "on the fly" at the RIW, or, if no specific format for data collection is considered.

In the "Prepare for Risk Expression" subsection of this chapter, I convey the logic and sense of urgency related to utilization of the REI format for collecting risk descriptions. Later in this chapter, I will give a specific example of an RR format and of plots and tables that can be used to convey results from risk analyses. The point I want to make here is to have established—prior to the RIW—exactly what forms will be employed and the format of those forms.

A veritable cornucopia of reasons exists that support such preparation. First, forethought and foresight makes it look like you know what you are doing—even if that is not so much true. Senior staff are not easily duped, and if they are to have confidence in your ability to lead them, then every effort should be made—prior to the meeting—to prepare. Look and act like an expert.

Secondly, there likely will be no "do overs" with regard to data collection. If you convene a multiday RIW and later discover that you have failed to collect one or more critical pieces of information, attendees of the meeting are not likely to appreciate having to revisit the risk information just so you can ameliorate the situation. This makes you look like an idiot and goes a long way toward eroding the organization's confidence in you and in the entire process.

Consistency is yet another paramount aspect impacted by preparation. At the RIW and at other times and meetings, a multitude of groups and individuals will participate. These independently functioning entities will each be collecting similar information and data related to the categories of risks they are addressing. If left to their own devices, such cadres of bodies will resort to a fixed method that they have devised for addressing the requisite tasks. The trouble is, each group will settle on a different methodology and, in the end, those diverse approaches might not lend themselves to easy confluence. Therefore, to achieve consistency in information and data collection, a small set of predetermined forms and formats is essential.

Thus far, I have addressed only the information/data-collection aspect. However, just as important are the forms and formats for conveying results from the risk process. There are at least two reasons to strive for consistency in reporting.

One rationale for report consistency is to avoid disappointment. For example, it should be agreed between at least the PM and the RPP—long before convening of the RIW—just what message will be conveyed to upper management as a result of the RIW. In addition, results from the RIW need to be distributed in a meaningful and practical manner to the disciplines that participated in the meeting. Believe me, even though it might be unspoken, the PM has in his/her mind an idea of what will result from the RIW. If these nascent concepts are not fully disclosed, discussed, and captured, the PM, at the very least, will be disappointed in the RIW results. In addition, the heads of the various disciplines—again, whether explicitly expressed or not—expect from the RIW some concrete and comprehensive recommendations upon which they can act. Make sure that the information that is conveyed to the various areas of expertise meet their expectations. The only practical way to achieve success in this area is to plan ahead.

A second primary reason for striving for consistency is to assure the comprehensive nature of the effort. Not too much is more demoralizing or downright embarrassing than to have completely missed a critical element of the analysis. I mentioned above that there are no "do overs" in this business, and preparation of the forms and formats—prior to their use—is one way to attempt to avoid missing a critical element. This lesson applies not only to information/data collection but also equally to message delivery. You are likely to get just "one shot" at delivering the results of the RIW to decision makers. Again, it makes you look like an idiot if you later discover that a critical element was not conveyed. Plan ahead!

Create Translation Table

To some, it might seem as though the subject to be addressed in this sub-section constitutes the discourse on monetization—it is anything but. I have made mention of the translation-table approach in Chapter 5 section "Poor Translation of Qualitative Data into Quantitative Data," subsection "Translation of Risks," but will fully address the subject here.

With regard to threat and opportunity impact, the coefficients collected in the RR should be in parameter units, which will be used to ultimately rank the risks. That is, just for example, if the PM and RPP have decided that risks will be ranked by their impact on NPV, then the consequence related to each risk should be recorded in the RR in units of NPV (millions of dollars, or the like).

Well, you say, how do I express in NPV my political risk of not getting our permits on time for a construction project? Funny you should ask, because I am going to address exactly that question in this section.

Prior to the RIW, the units in which the risks will be expressed and by which they will be ranked has to be decided. In this example, I have chosen NPV, but that is not to say that there are not many other parameters that would serve a given risk analysis better.

It is the duty of the RPP to sit down—usually with the commercial analyst (whoever is the keeper of the project economic model)—with the purpose of generating a simple translation table. For our hypothetical project, most risks will impact

1. Time (schedule)
2. CAPEX (capital expenditures)
3. OPEX (operating expenditures)

The RPP and the commercial analyst have to have some idea about the range of time and costs regarding the project. For example, if the project—and allow me this ridiculous example—was to buy a box of pencils, it would make little sense to assess time impacts of years or costs in the range of millions of dollars. Just as obvious, on a major construction project, it makes little sense to estimate impacts for minutes of time or tens of dollars.

Together, the RPP and the keeper of the economic model should decide upon a small number of categories for, say, schedule slips (either way—advancing or retarding the schedule). So, for the major construction project being considered, the RPP and the commercial analyst decide that they will translate into NPV delays in the following ranges:

1 month to 3 months

3 months to 6 months

6 months to 9 months

9 months to 12 months

12 months to 15 months

They have decided that if any discipline or risk delays the project for more than 15 months, then top management has to decide whether or not the project should proceed. So, the economic spreadsheet already contains numbers that represent the base case for the project (and remember, the base case is likely to already have accounted for most opportunities—even those with relatively low associated probabilities). The NPV that results from the base case is the starting point.

Now, the commercial analyst enters a delay in the project—usually a delay in obtaining the first commercial output from the project—of 3 months. The NPV associated with this delay is recorded. Next, 6-month, 9-month, 12-month, and 15-month delays are entered and the resulting NPVs are likewise recorded. This results in a set of NPVs that, between which, linear interpolation can be made.

At the RIW, the keeper of the economic model might supply each high-level group with a computer that has loaded on it the economic model. A separate worksheet can be devised on which there are just two or three input cells and a single output cell. In two of the input cells, the RIW members might enter one of two impact values associated with the risk (one cell for time impact and one cell for money impact). The third cell can be used to input a deterministic probability associated with the risk. RIW participants need only to then push the "go" button to see the "translated" NPV value (no accounting for the time value of money is done at the RIW—that is a bridge too far). In this process, the probabilities are used as "traffic cops" in the economic model. If such computers and models are not afforded the group, then members will have to combine the already-translated NPV values—as described in the previous paragraph—with the risk's probability by multiplication. This is not preferred, but it is sometimes necessary. What we are seeking, in the end, is a probability-impacted common metric—NPV in this case—so that ranking of the risks is possible.

During the actual risk-monetization process that will be described in Chapter 8, when a risk's delays are entered into the spreadsheet model, the risk's probability also should be entered. The probabilities will be used in "traffic cop" mode (see section "Decision Trees Applied to Legal Matters" of Chapter 4). When probabilities are applied, a risk impacts the base-case NPV only the percentage of times indicated by the probability percent. When this risk does not impact the NPV, the base-case NPV remains unchanged.

Costs are treated similarly. Prior to the RIW, the RPP and the keeper of the economic spreadsheet should deduce what few ranges in, say, CAPEX will be used. They might decide

$100,000 to $2,000,000

$2,000,000 to $4,000,000

$4,000,000 to $6,000,000

$6,000,000 to $8,000,000

Again, like with time delays, they have decided that if the CAPEX cost goes up by more than $8,000,000 for any given risk, then the entire project has to be reexamined by top decision makers. Just as with schedule delays, each cost-category limit will be run through the financial model to determine its impact on NPV. The NPVs of those values ($2,000,000, $4,000,000, and so on) are recorded. NPV impacts associated with any given risk can be linearly translated between the recorded values. Probabilities are applied as described above for schedule impacts.

It should be noted that such impact ranges should also be set up for opportunities. That is, NPVs should be generated for time and cost SAVINGS.

Now, I can imagine what you are thinking. In your mind, you're saying, "Glenn, NPV is the result of a time series of discounted cash flows and you have not accounted for the timing of the delays or the spends in the method you outline here!" If you were thinking that, you would be right. This is exactly why this step does NOT constitute risk monetization. The method outlined above is only meant to give us a rough estimate of the impact of each risk so that we can rank the risks relative to one another. The importance of the time value of money is not lost on me, but it is too cumbersome an element to deal with when trying to plow through, potentially, the evaluation of hundreds of risks in a couple of days. Asking for the relative times associated with the costs, for example, is just a bridge too far in the quick-paced RIW process. Some things you just have to learn to live with.

I believe that the creation of the translation table, as described above, is the most practical method for the relative evaluation and ranking of risks (note that I did not say that it was the best method—which would include consideration of the time value of money). However, there are myriad other risk-ranking methods commonly employed.

Again, in Chapter 5 section "Poor Translation of Qualitative Data into Quantitative Data," subsection "Translation of Risks," I briefly describe the approach to ranking risks—and this is the one that is likely most used—of relating categories of costs or time delays/advancements to a quantitative representation such as High, Medium, Low. The reader is referred to the aforementioned section of Chapter 5 for more detail. Ranking of risks using this approach typically comes down to sorting the risks such that those with the greatest number of "High" designations come to the top of the list. Of course, parameters that have High, Medium, and Low designations for threats or opportunities can be weighted so that a "High" related to one potential ranking parameter does not have the same "weight" as a "High" for a different potential ranking parameter.

I would go into detail about these ranking schemes and show examples of them if it were not for the fact that there are just so many variations on the theme that illustrating any one or few might actually do harm. Readers might take the illustrated examples as suggestions regarding how risks should be ranked for their project. Projects are so diverse that it is folly to try to give representative or generic ranking schemes. The translation table approach described in this section is the best example of a generic ranking scheme that I could propose.

Some Final Suggestions

With regard to preparing for the RIW, I have some final suggestions. Some of these have addressed in previous sections and chapters of this book, but those that have been mentioned before bear mentioning again for the sake of emphasis.

In the subsection of this chapter entitled "Can Call Off Critical Meetings," I indicate that the RPP should be empowered to call off the RIW if attendance is poor, if substitutes for critical attendees show up, or for a host of other reasons. Don't be afraid to utilize this power. It is a tough thing to do, when you are standing in front of a room full of people, each of whom has set aside the days for the meeting, to announce that the meeting is not going to happen and that it will be rescheduled for a future date. This also is likely to not go over well with the PM and others. However, it is far better to suffer the delay and the associated "slings and arrows" of postponement than to spend two or more days conducting a meeting that will yield poor results because of inadequate attendance or because of the lack of required experience of the attendees.

Also, be prepared to take a position of leadership and authority. Experienced hands in attendance can "smell" fear and lack of confidence. Taking a "cocky" attitude also is a mistake. The RPP should, however, start the meeting by exuding confidence and communicating the fact that he/she expects cooperation. A hesitant demeanor will result in a less effective engagement.

This final point might seem trivial, but it is supremely important. Prior to the RIW, make sure that the PM knows that he/she is expected to stay in the meeting from beginning to end. This cannot be implied by the RIW but must be expressly stated to the PM and an assurance of attendance related by the PM. It has been my experience that a PM will indicate, prior to the meeting, his/her intention to attend the full engagement only to be pulled out of the meeting (sometimes never to return) by more pressing issues.

It has to be agreed by the PM that there will be no more important issues while the RIW is convened. If the PM wanders off, you can expect that other attendees will also be "called out of the meeting" by "emergencies." If that happens, pretty soon the RPP can find that the attendance has fallen below critical mass and poor results will follow. If the PM stays through the meeting and expressly states that he/she expects others to do the same, the quality of the results is much enhanced.

Opening the RIW

Ah, now the fun begins! There you stand in front of a collection of discipline experts who likely don't know anything about you, don't really want to be there, don't actually understand the true purpose of the workshop, think they have better things to do, and feel a bit threatened by the entire risk process. Man, things just couldn't be better! So, how to start?

Review Again the Importance and Aim of the RIW

First, be enthusiastic. If the RPP and PM exude an air of hesitance or trepidation, things likely will not go well. Smile. At least act excited about the upcoming adventure. Speak slowly and deliberately without a preponderance of "ums" and "ers" and other speech interrupters. Make eye contact. Know what you are going to say, but don't—for goodness sake—don't read what you are saying! You can't speak with authority if you are reading from a script. Confidence begets confidence. Enthusiasm is met with enthusiasm. Unless you are a professional comedian, skip the jokes. Jocularity is marginally beneficial when it works and is absolutely deadly when it doesn't!

In the "Appeal to Their Sense of Logic," the "Appeal to Their Egos," the "Be Ready with Stories of Success," "Understand the Workflows," and Understand the Incentive System" sections of this chapter, I describe some tried and true methods that an RPP can use to instill a sense of urgency, purpose, and importance in organizational personnel regarding the RIW. This is the place where the RPP puts those methods to use.

Whether expounded by the RPP or the PM, at the opening of the RIW, it should be made clear that, for example, foreseeing opportunities and threats and doing something about them leads to a better project. A better project leads to job security. In addition, a not-so-veiled appeal to their egos is not out of order. Make sure that they realize that they—as representatives of their disciplines—are critical to success. An expression of how the risk process will link with their typical workflows and how adherence to the risk process will meld with their incentive system is a good idea. If there are previous examples of how implementation of a risk process—in their business or in another area—had a significant positive impact on a project, now is the time to relate it. The bottom line is that the population in the room should be impressed, to the extent possible, with the importance and aim of the RIW and, subsequently, the entire risk process.

Address "What Is the Question?"

In the "Be Clear as to 'What is the Question'" section of Chapter 3 and in the "Make Sure that the RPP and PM Agree on 'What Is the Question'" section of this chapter, I stress the importance of the two parties coming to agreement regarding just what problem is to be addressed by the project. I will not repeat that advice here, but at the opening of the RIW, it is critical that all parties hear from the PM just exactly what problem that is to be considered in the RIW.

Remember, each person in the room will have an individual "understanding" of what constitutes the project to be addressed at the RIW. Time should be set aside in the opening of the RIW to explicitly state the scope and definition of the project and to have an open discussion—including questions and answers—of the project scope and purpose. Everyone in the room must be "on the same page" before risks are identified, described, and assessed. To the extent that you can, be assured that a common understanding is held by all.

Emphasize Opportunities

Just as important is to convey that a risk process is every bit as much about identifying opportunities as it is about describing threats. Many of the people in the room will be of the mindset that "risk" means "threat" or "downside."

Regardless of how much or how little training the participants might have had prior to showing up at the RIW, make sure that the RPP emphasizes that the success of any project—and he/she should make specific reference to current-project-related examples—is just as dependent upon identifying and capturing opportunities as it is upon similarly treating threats. If this point is not hammered home, the list of risks emanating from the cadre of discipline experts will be a litany of threats. This is really an important point.

Review How to Express a Risk

I address the actual process of describing threats and opportunities in the "Risk Description—How to Express a Risk" section of Chapter 2 and in the "Prepare for Risk Expression" section of this chapter. At the opening of the RIW, it is time to put that learning into practice.

Regardless of whether the RIW participants have had previous training in how to express a risk in REI format, and in addition to having the 3×5 risk-identification cards preprinted with the REI format, take time at the opening of the RIW to make example of how this format can be used to describe risks related to the project being considered. I know that this advice sounds academic, but I can tell you from many years of personal experience that if risks are properly expressed at the beginning stages of the RIW, it saves a world of trouble and pain in later RIW steps. So, make it a point to take time to convey examples of poorly expressed and correctly expressed threats and opportunities that relate to the project.

Review What Probability Means

In the "Probability of Occurrence" section of Chapter 2 and in the "Uncertainty and the Confusion that Surrounds It" section of Chapter 3, I describe in detail the many aspects of probability. At the beginning of the

RIW, the RPP should not launch into a diatribe regarding the technical aspects of the meaning or application of the term "probability." However, the subject of probability has to be addressed. Believe me, if you don't make an issue of probability at the beginning of the RIW, then many of the numbers collected at the workshop will be worthless.

Personally, I always start with simple descriptions of what "probability" is not. For example, probability is not frequency, or how many times something has happened in the past. Probability, unlike frequency, is a projection of the likelihood of occurrence of the issue in the future. Probability also is not the likelihood that we are going to be able to do something about the risk (mitigate a threat or capture an opportunity). You might be surprised at the number of people, when asked about the probability associated with a risk, are thinking about the likelihood that they can "fix" it. This is *very* common and needs to be avoided.

Following a discourse on what probability is not, I launch into what is expected as an expression of probability. I relate that we are asking for their opinion—usually expressed as a range—that the issue being considered will materialize on *this* project *in the timeframe being considered.* For example, we are not interested in the probability of a car accident occurring in the city as a whole if our project is addressing just one street. Also, we are not interested in the probability of an accident ever occurring—but only in the timeframe considered by the project. This is just one reason why it is *so* important to have the RPP and PM describe the scope of the project at the beginning of the RIW. If everyone in the room is mentally considering a different scope and definition of the project, imagine how useless will be the expressions of probability. Make sure this issue is addressed at the outset of the RIW.

In addition, be sure to make clear that probabilities are to be expressed in units of percent. Again, you might be surprised at the number of individuals who wish to express probability as "1 in 7" or "15 out of 35" or some other equally useless description. Demonstrate by example what is expected.

Review What Impact Means

Again, in Chapter 2 and elsewhere in this book, I have taken up the issue of impact (or consequence). Most of those previous discourses were attempts by me to describe to you, the reader, the definition, format, handling, and so on of impact. As with the other topics addressed in this section, it is at the outset of the RIW that the RPP must be assured that all attendees share a common comprehension of the concept of impact.

First, it should be made clear that impact is separate and distinct from probability. A risk can have an associated high probability along with either a high or low impact. The reverse also is true—a low probability paired with either a low or high impact.

Impact, like probability, is to be considered in the scope and timeframe of the project. In addition, impact is to be first expressed as a range and in the natural units. For example, a schedule advancement (an opportunity) should first be expressed in the relevant "time chunks"—days or weeks or months or whatever. Subsequently, the impact should be translated—using the translation table previously discussed—into a universal expression of impact. An example might be NPV. In this way, impacts originally expressed in, say, weeks or tons, or CAPEX or numbers of personnel, etc. can be compared and ranked by a common measure. Be as sure as you can be that every member of the group understands this process.

In addition, make it clear that each impact expressed does not reflect the cost of mitigation/capture. In the risk process, the cost of mitigation of a threat or capture of an opportunity will need to be compared against the impact ("cost/benefit" analysis). This subject is covered in some detail in the "Cost of Threat-Mitigation or Opportunity-Capture Action" section of Chapter 2. If the expression of impact already includes the cost of mitigation, for example, it is impossible to genuinely generate a cost/benefit estimate. Backing out the mitigation or capture cost at a later date from the expression of impact is difficult, cumbersome, time consuming, and usually ineffective.

Review the Meaning of Mitigation and Capture

As mentioned in the immediately preceding section, costs associated with the mitigation of each threat and with the capture of each opportunity will be recorded. It is essential that when expressing those costs, each attendee have in mind the same concept.

Even when organizational personnel have had training in the fundamentals of risk, it will be true that when most of those people think about, for example, mitigation, they will be thinking about how much it will cost to "fix" the problem if it occurs. It seems that this is a more natural way to consider the cost of mitigation (or, opportunity capture).

For example, if you were to be asked what would be the mitigation cost of a broken transmission on your car, your mind might go immediately to the cost of fixing or replacing the transmission. Costs associated with this mindset are not what we are trying to capture.

In the risk world, the term "mitigation" for threats or "capture" for opportunities indicates the costs we might incur when we take steps to prevent the threat from happening in the first place (or, to assure capture of the opportunity). Part of the point of a risk process—and a main point to the RIW—is to foresee problems or opportunities that might present themselves within the scope and timeframe of the project. If we have, for example, identified a probable threat, then we need to consider what steps we would take, and the cost of those steps, to try to ensure that the threat does not materialize.

I will fall back on my tried-and-true example of the car trip. Before the trip, we discover that some of the lug nuts on one of the wheels are loose. This is a threat. If we take the trip and the wheel comes off while we are driving, the cost of a new tire, new wheel, and any resulting body damage could be significant. This is the cost if "fixing" the threat after it has materialized. This is the cost that most people naturally consider (the cost of the broken transmission, as I mentioned above).

What we actually want, however, is to capture the cost of the mitigation action. That action is the one, having foreseen the possible threat, that we take to try to ensure that the threat does not materialize. In the case of the loose lug nuts, it would be the cost of taking the car to a garage and having the lug nuts tightened. This cost is significantly less, in most cases, than the cost of "fixing" the consequences of the threat after it has happened. Be sure that everyone understands the meaning of "mitigation" and of "capture" and the associated costs to be captured.

Start Identifying Risks

Now we come down to the meat of the matter—identifying the risks. This topic has been directly and tangentially addressed in various preceding chapters and subsections. In this section, I will offer practical advice regarding just how to carry out the risk-identification process in a real-life situation.

Capturing the Risks

Few of the activities in the RIW are performed utilizing the entire group of attendees. Identification of risks is one exception. In the "Determine Main Risk Categories and Who Will Address Them" section of this chapter, I delineate how high-level risk categories should be determined prior to the RIW. In my software design/building example (see previous section entitled Determine Main Risk Categories and Who Will Address Them), I suggest the following categories:

- Location and Business Type
- Commercial/Competition
- Design/Engineering
- Information Technology Innovation
- Installation/Maintenance

In the "Prepare Forms" section of this chapter and in various other sections of this book, I address the expression of risks in REI format and in the "Prepare Forms" section, I advise that preformatted, "sticky-backed" 3×5 cards of red (for threats) and green (for opportunities) be utilized.

Red and green preformatted (each card has a Reason, Event, and Impact section) cards are handed out to each RIW participant. It is requested that each person think of as many threats and opportunities as practical, record those risks on the appropriate card in the required format, and then stick each card on a board or on a large sheet of paper with the appropriate heading. This process typically consumes about one-half hour to one hour to complete. In the end, each high-level category will have, usually, dozens of red and green cards associated with it.

This process, admittedly, is the low-tech approach. There are things about it that I like and things that I don't care for, but overall, the benefits, in my opinion, outweigh the drawbacks.

One benefit of this low-tech method is that it allows each participant to think broadly about the project—that is, think about categories that are not necessarily in their domain of expertise—and anonymously post risks that they perceive. This process yields expressions of risks that are more numerous, honest, and that emanate from many perspectives.

Contrast this method with the more high-tech tactic of, for example, projecting a spreadsheet on a screen (using a laptop and a projector) and then asking the audience to, one at a time, "shout out" risks for any of the major categories. This process, you will find, stifles creativity and contribution. Nonexperts in a particular realm are reticent to express their opinions and viewpoints and the resulting list of risks is relatively sterile.

One drawback to the low-tech approach is that it is inevitable that the same risk will be repeated multiple times by various attendees but expressed slightly differently. Because this is true, one of the main tasks of the subgroups that will address the high-level categories is to consolidate the expressions of risk. This takes time and patience. Contrast this process with the projected-spreadsheet approach in which risks are "shouted out" one at a time. Typically during the course of this approach, duplicate risks can be caught and not repeated.

Another drawback to the low-tech approach is the inevitable "swapping" of risks between high-level categories. After the plenary session in which risks are recorded and posted under one of the high-level categories, the group is broken up into subgroups—one for each high-level category. It is very typical that, after reading a risk that was originally posted under, say, the Commercial/Competition heading should actually be addressed by the Installation/Maintenance group. Using the spreadsheet/projector method, each risk can be discussed as it comes along and a decision made regarding to what high-level category it belongs.

Most of the arguments made above might seem to favor the higher-tech approach. Though that might be true, I remain a defender of the low-tech method because I find that the preponderance of significant risks are those identified by people who are not experts in the area of one or more of the high-level categories. The richness of the risks expressed in the higher-tech plenary session pales in comparison to the scope and breadth of risks that emanate from the anonymous-contribution low-tech approach. If the salient point of any risk-identification process is to capture the broadest and most insightful set of risks—and it is—then the low-tech method will prevail every time.

Break-up into Preassigned Groups

In the "Determine Main Risk Categories and Who Will Address Them" section of this chapter, I discuss how you might populate each of the high-level category groups. The reader is referred to the aforementioned section. Having decided prior to the RIW, which individuals will be assigned to each high-level-category group, it is now time to actually do it and get to work.

Each group should have a leader. I find it best to let the group select its own leader with the caveat that the leader can't be someone who is an expert in the category area. For example, the leader of the Commercial/Competition subgroup should not be the commercial analyst in the room. Folks who are experts in the area tend to have strong preconceived notions about how the work of the group should be done. Nonexperts will take a much more open-minded (stemming from relative ignorance) approach to the subject. You might perceive this as a drawback but, believe me, it is a benefit.

First on the list of tasks is for the participants, as a group, to read each expressed threat and opportunity and to attempt to identify those risks that have been repeated multiple times—expressed a bit differently each time. In the "Correlations between Threats and Opportunities" section of Chapter 2, I give several examples of this phenomenon. Having identified the multiple expressions of, basically, the same risk, the group should rewrite the risk so as to capture the essence of each of the "repeats." This consolidated risk replaces all of the original multiple expressions of the risk.

Also, as pointed out in the "Correlations between Threats and Opportunities" section of Chapter 2, sometimes the same risk is expressed as both a threat and an opportunity. For example, a risk expressed as a threat might be

> If the contracts are not signed before equipment is ready to be delivered, then equipment will have to be stored at an alternate site, causing a significant impact in CAPEX for the project and a delay in the project schedule.

Alternatively, the same recognized issue that relates contract signing and equipment delivery might be expressed as an opportunity.

> If the current rapid and favorable pace of contract negotiations continues and contracts are sighed in a timely manner, then equipment could be delivered to the construction site early, offering a significant savings in CAPEX and, more importantly, an advance in the project schedule.

Clearly, these two expressions pertain to the same situation. It should be resolved by the high-level-category group whether the risk is best expressed as a threat or an opportunity and members of the group should restate the risk in proper REI format.

Record Risks in the Risk Register

Regardless of whether the project RR will exist as an on-line and Web-based application, a simple spreadsheet, or some other manifestation, each high-level-category group will be supplied with a copy of the blank (but already formatted) RR. It is best if the RR can be projected onto a screen for all to see.

When the group has completed the tasks of consolidating and restating all of the risks in their particular high-level category, it is time to record those risks in the RR. There are several pertinent steps.

In the major section "Risk Registers and Their Many Implications" of Chapter 2, I give an example of a stylized RR. As I say in Chapter 2, the RR shown in this book is by no means meant to be a comprehensive example. Each project is different and will require a RR that specifically suits. However, there are a few parameters that likely will be common to most registers.

In Chapter 2, I describe the Risk Identifier, the Short Risk Description, and the Risk Description. The reader is referred to those discourses. Again, having completed the tasks of consolidating and restating all of the risks in their particular high-level category, the group should set about the task of establishing a Risk Identifier and a Short Risk Description for each risk and record those items along with the complete Risk Description in the preformatted RR supplied.

If performed properly and completely, the tasks described above—starting with the introductions by the RPP and PM, the review of critical elements, the risk-identification session, and the activities so-far described by the high-level-category groups—will fill the first day of the RIW. It is recommended that the leader of the RIW strive to get this far in the process in day-1 of the RIW. The activities of the second day are delineated below.

Assessing, Valuing, and Ranking Risks Prior to Monetization

An RIW will typically result in the identification of, literally, dozens to hundreds of risks. This represents a collection of threats and opportunities far too voluminous to even dream of addressing all risks in a monetization process. Therefore, an initial culling or ranking of the risks—based on their relative "importance" to the project—must be performed. A monetization process is applied only to those relatively few risks that "bubble to the top."

Methods utilized to relatively and absolutely rank risks range from the inanely simple to the ridiculously complex. In the next sections, I will relate just a few of the more popular methods employed.

"Just-Shout-It-Out!" Approach

First, it should be established whether there is going to be just one assignment (H, M, or L) for the risk, or, whether there will be separate designations for impact and probability (see Figures 6.1 and 6.2). If just one assignment is to be utilized, then a definition for the classification must be established. For example, does the H/M/L translate to overall "importance" to the project or does the label designate impact only, or will participants be asked to mentally combine probability and impact prior to "shouting out" their opinion on H, M, or L?

If the single-tag (H, M, or L) process is utilized, then the initial ranking of the risks is complete. Likely all of the risks designated "H" will be addressed somehow and probably a cadre of those risks sporting labels of "M" also will get further treatment. A simple columnar "ranking" of the risks is the quintessential representation of risk prioritization.

An extension of this method of assessing risks is to assign a separate H, M, and L to probability and to impact. So, a given risk might be determined to have an L (low) probability of materializing, but an M (medium) impact if it does occur. Again, the designation of H, M, or L for either parameter (probability or consequence) is arrived at by consensus opinion and, therefore, is qualitative in nature. Given that estimates of both probability and impact are made, risks can be plotted in probability/impact space as shown in Figure 6.1. From observation of the absolute and relative positioning of risks on the plot, determinations can be made regarding which risks will be further treated. Further analysis regarding what can be done about a risk (ability to mitigate or capture) can be assessed and plotted as shown in Figure 6.2.

I would advise that the process described above be eschewed if possible. Truly, it is simple and expedient, but if at the end of the day you are going to be expected to defend the dispensing of CAPEX to mitigate a risk or to capture an opportunity, or, if you are expected to give evidence regarding the benefit to the project if the top-ranked risks are addressed, then quantitative data ultimately will have to be collected. Returning to each risk at a later date for further examination is cumbersome, time-consuming, and organizationally impractical.

Quantitative Approach

A "Just Shout It Out!" qualitative approach to defining, valuing, and ranking risks defines one end of the methodology spectrum. The opposite end of the method scale is exemplified by quantification of nearly every parameter associated with a threat or opportunity. This is the method I prefer and, therefore, recommend.

Collect Impact Data

One step in what I would consider to be the right direction is to collect, at the RIW (probably second day), quantitative data related to the impact of the risk. There are three issues to be addressed. First is that of the units in which the data will be recorded. Second is whether to collect the data as discrete values or as a range. Third is the consideration of a secondary impact.

With regard to units, collection of impact in "natural" units is best. That is, if a delay related to a risk is most naturally thought of in seconds rather than in minutes, hours, days, months, or years, then collect the impact data in seconds. Mistakes are made when participants are asked to evaluate risks in unnatural units.

For example, if I were to ask you about your salary, if you are an hourly employee, you might feel best about expressing your salary in dollars/hour. If you are a salaried employee, then you likely are most comfortable telling me how much you make per year (dollars per annum). It might be that other time-related impacts in the project analysis are being related to months or days. Just because "days" are used with other schedule-impacting risks, it is a mistake to ask either the hourly or the salaried employee how much they make per day. It is far better to allow the employee to express their salary in what they consider to be "natural" units and to use a simple calculation to later convert that expression into the desired time unit rather than to attempt to collect the data from the employee in units with which they are unfamiliar.

With regard to the second issue—that of collecting impact data as discrete values or as a range—I would advise you to (almost) always strive to use a range to represent impact. Clearly, this has its limitations.

For example, if the risk model is concerned with the magnitude of a person's salary at the present time (not considering whether the people being evaluated might get promotions, raises, bonuses, salary reductions, etc.), then it is silly to attempt to collect a discrete value as a range. Each person is sure of what is their salary today, and if that is what the model requires, then no uncertainty is involved—rendering use of ranges (distributions) moot.

Situations as the one described above are not typical, however. It is far more common to be uncertain about impacts. For example, one of the parameters in the project model might be the cost of fuel over a given time period. Just take a look at the sign in front of your local gas (petrol) station on any two consecutive days if you need to be convinced that fuel prices fluctuate. If the model requires an estimate of fuel prices over the entire time period of the construction phase of the project, then the range for fuel prices could be quite broad. In all cases in which uncertainty of impact is even possible, I urge the collection of impact data as a range.

In the "Probability of Occurrence" section of Chapter 2 and in the "Expressing Uncertainty" section of Chapter 3, I cover the subject of distributions in some detail—although not as thoroughly as I have in previous books. I also address in the aforementioned sections interview methods useful for gleaning distribution-building data from interviewees. The reader is referred to the sections of this book and to previous books I have penned for more details on this subject. These methods should be employed at the RIW to extract from the group rational ranges that represent the impact of a risk.

When utilizing the methods delineated in the aforementioned sections of Chapters 2 and 3, the range of impact should represent the entire range of possibilities—including those values that might cause abject failure of the project. For example, it might be true that when we drill a water well, we will require that the rocks from which the water emanates in the subsurface have at least 10% porosity (i.e., the rocks are comprised of 10% connected pore space that can hold water). We know from experience that some of the wells we have drilled in the past have had porosities down to 2%. Those wells with less than 10% porosity, consequently, had waterflow rates too low to allow them to act as successful and economic water wells. Those wells were abject failures due to less-than-required porosity values.

When collecting the data for the range of porosity for the proposed well (the project is to drill a new water well), the range should include those porosities—down to the lowest porosity previously observed of 2%—that caused abject failure of the previous wells. Those types of wells are "out there waiting for you" and it is possible that the new well will be one of them. The range of failure porosities (2% to just less than 10%) will be used, as will be described in a subsequent section, to determine the chance of abject failure for the project due to porosity combined with all other individual-parameter chance-of-failure (COF) values.

A third element to consider at this stage is the possibility that there might exist multiple impacts for a risk. For example, selection of one contractor over another might have direct cost implications—simply because one contractor charges more per hour than another—and might also have schedule impact because one contractor simply does not have the capacity to complete the job in as timely a manner relative to the alternate contractor. In this case, if the costs and time delays are independent, then there might be reason to consider multiple impacts for a risk. Although—mainly for space-on-the-page reasons—I will not in this book show any examples of multiple impacts for a risk, it should be noted that any RR or spreadsheet or other risk-recording vehicle should accommodate the recording of multiple impacts and probabilities for any risk.

Probability Weighting of Impacts

If you are a user of the infamous Boston Square (see "Use of the 'Boston Square'" section of Chapter 5), then you are a proponent of using probability to modify impact. Typically, the Y-axis of the Boston Square is defined as "Probability × Impact."

I, however, am a proponent of not mixing probability and raw impacts. In previous sections of this book, I have given good reasons to avoid this practice. Just a few of those reasons are

- Multiplying, for example, a low probability by a high impact results in a risk that plots in the middle section of the Boston Square. It is likely that you would want to be alerted to any risk that can have a significant consequence.

- The *X*-axis of a Boston Square typically has a label that reflects the ease or difficulty in "mitigating" the risk. For the definition of action items, it is very important to know whether a mitigation (for a threat or, a capture action for an opportunity) is aimed at affecting the probability or the impact. On a Boston Square, you can't discern this.
- Making Probability/Impact, Probability/Mitigation-Capture, and Impact/Mitigation-Capture plots are far superior to the Boston Square. Examples of these plots are shown in the "Deal with Them" subsection of this chapter.
- When attempting to estimate mitigation or capture costs and a cost/benefit estimate for a risk, the costs and cost/benefit ratio need to be generated relative to the unadulterated (not impacted by probability) impact of the risk.

For these and other reasons, I would advise that if one is tempted to utilize probability as an impact modifier for raw data that are entered into the RR, you record both the initial impact and probability values as well as the combined (probability-weighted) values. This is not the same as modifying impacts by probabilities to translate those impacts into a probability-affected common metric such as NPV.

Collect Probability Data

In the "Review What Probability Means" section of this chapter, the "Probability of Occurrence" section of Chapter 2, and the "Uncertainty and the Confusion that Surrounds It" section of Chapter 3, I treat in some detail the meaning and use of probability (i.e., it does not mean "frequency" and it does not mean the likelihood that we will be able to mitigate a threat or capture an opportunity, etc.). At this stage of the RIW, it is time to put those leanings to good use.

With regard to collecting probability as discrete values or as a range, I truly am of two minds. While my statistical nature prods me to advise you to collect a range for the probability of occurrence associated with any given risk, my practical bent lures me toward suggesting that the deterministic (single-valued) method is best.

Rationale behind representing probabilities as ranges—using distributions—stems only from the fact that such ranges might be used in a Monte Carlo setting. This is true enough, and some arithmetic problems can stem from representing probability as a range, but the most vexing conundrum is that people simply don't have a natural ability to think accurately and precisely about probability.

Contrast this with impact. If in an RIW the question of the range of the cost of a given compressor of a specific horsepower were to come up, numerous people in the room could bracket (i.e., generate a range for) the costs with significant confidence. However, if, as we define it, probability is not frequency (of occurrence in the past) and is an estimate of, in this case, the availability of a given compressor size at a specific point-in-time in the future, it is likely that estimates will be "all over the map." Getting a very broad range around such a probability—around a thing that nobody really knows about—is fundamentally different from obtaining as broad a range as possible related to a parameter with which members of the group have some experience (the cost of the compressor, for example).

Translation of the risk problem to a decision tree is just one reason to avoid ranges. If at some point in a risk analysis, someone wishes to represent the problem in decision-tree format, the probabilities associated with, say, three branches have to sum to 100%. If each of the three probabilities are represented by ranges and the decision tree is then transformed into a Monte Carlo model, it could easily end up that the sum of the probabilities is less than or greater than 100%. See Chapter 1—"Two Approaches to Solving Decision Trees—Class-Action Suit Example" in *Risk Modeling for Determining Value and Decision Making* by Glenn Koller (listed in the Reference section at the end of this chapter) for a detailed example.

I know the argument can be made that one could collect reasons why the compressor in question might or might not be available at the specified time in the future (as I recommend doing when collecting data for ranges—see the "Probability of Occurrence" section of Chapter 2 and the "Expressing Uncertainty" section of Chapter 3) and avoid the problem of anchoring. However, I find it sometimes—but not always—more useful in the RIW setting to come to consensus regarding a deterministic estimate of probability. I leave it to the reader and the RPP of the project to decide how probabilities should be represented.

Use Translation Table to Convert to Common Measure

Now, assuming that you did not promote the route of "Just Shout It Out" as described in a preceding section, representatives in the high-level-category groups will have collected either discrete or range-related data for each impact for every risk. Note that in a previous paragraph, I advised that impacts be collected and recorded in their "natural" units. If that is done—and it should be—then it is nearly impossible, and certainly impractical, to compare the impacts of various risks. For example—and this is a rhetorical question—how would you rank a 2-week delay relative to a $10,000 CAPEX increase? Hmmm … Something has to be done.

In the subsection of this chapter entitled "Create Translation Table" I fully address how such a table might be generated. Now is the time to put it to use.

Practical application of the translation table can be performed in at least three ways. The first method is to have the members of the high-level-category group calculate the translation of the "natural units" for a parameter's impact into the common unit to be used for risk ranking (a unit such as

NPV). Clearly, this takes precious time in the RIW and is prone to calculation errors. However, if a precalculated translation table is supplied to the
participants (see "Create Translation Table" section of this chapter), then it
is a fairly simple task to select the translated NPV from the table and multiply that NPV by the risk's probability. I don't generally advocate the use of
probabilities as multipliers, but sometimes it is the practical means.

A second method of transforming the impact of each risk into a common metric is to have the RPP, and probably some helpers, translate the "natural units"
for each risk into the common unit. This has both advantages and drawbacks.
One advantage is that it gives the RPP a chance to review the risk impacts and
to spot trouble before it gets any further in the risk process. The downside of
this method is that the RPP would have to perform this task in the evening
and night between the two RIW sessions. Pure physical exhaustion becomes
an issue as does the fact that if such impact numbers are to be had on the night
between the two RIW sessions, then generation of such numbers have to be
shoehorned into the first day of the session. This, most times, is not practical.

A third method—and the one that is recommended if it can be arranged—
is to supply each high-level group with a computer that has on it the economic
model. A separate worksheet can be put at the front of the economic model so
that the participants can simply enter a deterministic value for impact (time,
money, etc.) and the risk's associated probability. Pushing the "go" button
generates a deterministic NPV. In this process, the time value of money is
not considered, but the probabilities can be used in the program as "traffic
cops" rather than as multipliers and a relatively crude estimate of probability-
impacted NPV can result. This common metric, then, is used to rank the risks.

Regardless of the method chosen, the use of the translation table is absolutely
the preferred method of applying a universal metric (such as NPV) to all risks.
This metric can now be used to roughly rank risks relative to one another—
or can it? What about probability? Also, if we have translated the risks into a
common metric such as NPV, why does this not constitute the monetization
process? I'll address both of these issues in subsequent sections and chapters.

Collect a "Cutoff" Value

In the "Chance of Failure" section of Chapter 3, I address the subject of
selecting cutoff values for distributions and how those cutoff values are used
to define the COF value for an individual parameter, how the individual-
parameter COFs are integrated to derive a total chance of success (TCOS) for
a project, and how that TCOS is reflected in the "answer variable" cumula-
tive-frequency-curve plot. The reader is referred to the aforementioned section of Chapter 3 for a complete review of the subject.

If it is the intent that the risks captured in the RR are ultimately going to
be used in a monetization process and if the impacts and/or probabilities
for the risks have been collected as ranges, then a cutoff value for each range
should be considered and recorded in the RR. As delineated in Chapter 3,

these cutoff values will be instrumental in the Monte Carlo processing of individual-parameter distributions and will impact the cumulative frequency curves of calculated parameters. Such an example is shown, in principle, in the "Categories-of-Impact-and-Probability Approach" subsection of this chapter.

Identify Risks That Will Be Handled Separately

In the "Ignoring Soft Risks" subsection of Chapter 5, I discuss how risks that are of a sensitive nature—such as fatalities—might be handled. Such risks typically are not included in, for example, an economic-spreadsheet model or a subsequent Monte Carlo analysis. Other risks such as those that are low-probability/high-impact also need to be specially handled.

How to handle such risks is covered elsewhere in this book. The point of this section is to alert the reader to the fact that it is at this stage of the RIW when the members of the high-level-category groups and/or the RPP must identify those risks—as they put them into the RR—that will require separate or special handling. Such risks typically are not included in the RR and are recorded elsewhere and are separately addressed.

Categories-of-Impact-and-Probability Approach

Thus far, I have detailed two approaches to recording risk data in an RR. The "Just Shout It Out" method (no overt quantification of risks) and the "Quantitative Approach" involving collecting deterministic- or range-data on impacts and probabilities, creating translation tables, and numerous other steps detailed above. There is, however, a very popular middle-ground approach.

For this methodology, all of the advice given in previous sections of this chapter regarding risk identification, assigning a risk identifier, defining the risk at various levels, and so on still applies. The twist is in the quantification of the impacts and probabilities.

In this approach, preset ranges are established and each range is given an "index" value. For example, consider Tables 6.1 and 6.2. In Table 6.1 are listed three ranges of probability. Each range is assigned one of three index values: 1, 3, or 5. Similarly in Table 6.2, three ranges of impact are each assigned impact values of 1, 3, and 5.

Now, take a look at the stylized RR for a construction project in a foreign country shown in Figure 6.3. At the RIW, members of each high-level-category group are presented with Tables 6.1 and 6.2 and asked to classify the probability and impact for each risk into one of the three categories shown in the tables. The indices for each selected category are recorded.

For example, the first risk (threat) listed—related to a potential detrimental change in the host government—has been deemed by the Regional Considerations high-level-category group to have a Schedule Index of 5, a Cost/Revenue Index of 5, and a Probability Index of 5 according to Tables 6.1 and 6.2.

TABLE 6.1

Assignment of Percent Probability Ranges to Three
Deterministic Numerical Categories (1, 3, and 5)

Probability

5 = High (> 75%)
3 = Medium (>50% <75%)
1 = Low (<50%)

TABLE 6.2

Assignment of Consequence Ranges to Three
Deterministic Numerical Categories (1, 3, and 5)

Impact Revenue/ Index	Cost (+/−)	Schedule (+/−)
5 = Catastrophic	>$0.5B	>1 year
3 = Critical	>$0.1B <$0.5B	>0.5 year <1 year
1 = Marginal	<$0.1B	<0.5 year

Risk ID	Concise risk description	Risk description	Schedule index (+/−)	Revenue/cost index(+/−)	Probability index	Schedule severity (impact × probability)	Cost/revenue severity (impact × probability)	Mitigation action	By whom?	By when?	T = Threat, O = Opportunity
REG9	Government overthrow	If the current government is overthrown, contracts might have to be renegotiated causing schedule delays and an increase in CAPEX lowering NPV.	5	5	5	25	25				T
COM5	Early sales	If the plant is completed ahead of schedule, sales of product to local utilities will defeat competitors and increase revenue, enhancing NPV.	−5	−3	3	−15	−9				O

FIGURE 6.3
Risk ranking using index values from Tables 6.1 and 6.2.

If you are one of those people who just have to multiply the probability by the impact, then the product of Schedule Impact Index and Probability Index can be used to arrive at the value of 25 indicated as Schedule Sensitivity. Similarly, the Probability Index and the Cost/Revenue Index can generate a product of 25. This is shown as Cost/Revenue Sensitivity.

Opportunities can be similarly treated. The second risk listed in Figure 6.3 is an opportunity identified by the Commercial Considerations high-level-category group. Indices selected for the opportunity also are taken from Tables 6.1 and 6.2, but impact indices are recorded as negative values so that the product of impacts with probability will result in negative sensitivity coefficients. This is done simply to distinguish opportunities from threats. When ranking all risks, it is the absolute values of all risks that would be utilized in the ranking process.

Regardless of which of the three methods you might employ at the RIW to roughly evaluate and rank risks (or some twist on any of the three methods shown above), it is essential that all risks be somehow "valued" and ranked.

Mitigation/Capture Actions

Regardless of the method used to assess, value, and rank the risks identified in the RIW, each risk should have assigned to it attributes that address how the threat or opportunity will be mitigated/captured, costs associated with mitigation/capture, responsibilities and responsible parties, due dates, and the like. All of the items I will address briefly below are covered in detail in the "Risk Registers and Their Many Implications" section of Chapter 2. I mention these items again here because this is the point in the RIW at which these actions are carried out. Again, the reader is referred to the aforementioned section of Chapter 2 for details on the items addressed below.

For each risk, members of the high-level-category groups should

- State the mitigation/capture action that will be taken. This statement should contain no questions, dates, or costs/revenues.
- Record a probability of effectiveness of the mitigation/capture action. For example, if the mitigation action for a threat is to have prior-to-the-project meetings with government-agency officials to try to ensure that permits will be granted on time, the probability— expressed in percent—should reflect the likelihood that the meetings will have the intended positive impact. The probability can be recorded as a deterministic coefficient or as a range of values.
- State whether the action item is aimed at addressing the probability associated with the risk, the impact of the risk, or both.

- Estimate the cost of the mitigation/capture action. This can be expressed as a deterministic value or as a range. If estimating the cost of a capture action for an opportunity, this cost should not be offset by the benefit of capturing the opportunity.
- Record due dates for mitigation/capture actions.
- Record responsible parties for reporting progress to the RPP—the action owner. Make sure this party knows what information the action *taker* (different from the action *owner*) needs to collect to satisfy requirements that define when the action is complete.

A stylized RR that reflects the "Quantitative Approach" methodology is shown in Figure 6.4. The register shown is meant only to be exemplary and it should be recognized that any RR for a real-life project will likely have to be somewhat customized.

Some Post-RIW Activities

Following the RIW, there are a host of activities that should be carried out. Just a few of the tasks that are generic to almost all projects are listed here. Of course, a specific list of post-RIW tasks will be dictated by the type, scope, and so on of the project.

Declare Victory

Not that you should feign success or hide your real feelings concerning how went the RIW, a responsible RPP will at the very least prepare a message to all participants and their management regarding the RIW. This should be done within a day or two of the close of the RIW.

First, thank everyone—participants and their managers—by name. Be sure to be comprehensive and stress the importance of their participation.

Declaration of victory should be done at some level. Remember, just getting all of the participating parties gathered in one place is in itself a triumph. Highlights of the meeting including summary statistics such as the number of risks identified, the categories addressed, some of the most important risks defined, and one or two major action items should comprise the message. If a plethora of plots and charts are included in this general message, people will generally eschew reading the communication.

Share Detailed—but Summarized—Outcomes

It is a mistake to send around a copy of the final RR. However, it is beneficial to send to meeting participants a separate communication (separate

Project name: Full lifecycle widget-plant construction

Risk ID	Concise description	Risk description	Probability (%)	Impact units	Impact minimum	Impact peak	Impact maximum	Translated impact (NPV in $MM)	Mitigation/ capture action	Probability of action effectiveness (%)	Action cost	Action owner	Action implementer	Due date
COM 17	Government overthrow	Click here for full description	60	CAPEX in millions $ (MM)	1	3	9	−5	Early talks with opposition party. Reduce probability of bad relationship.	50	$0.5MM	Joe Smith	Jane Doe	1/1/20XX
TEC 21	New technology	Click here for full description	75	Weeks	20	35	45	4.5	Install demo plant technology. Increase economic impact.	90	$1.3MM	Sam Bar	Jack Trades	2/12/20XX

FIGURE 6.4

A stylized "quantitative approach" RR in which risks can be ranked by the product of the probability and impact category coefficients.

from the one described above in the "Declare Victory" section) that contains some detail regarding the risks.

It is customary to disseminate summaries of costs, the number of threats identified by category, the number of opportunities identified by category, the rough impact on NPV of the "summed" threats and opportunities, Impact versus Probability charts for each category and other pertinent-to-the-specific-project plots, tables, and charts.

Avoid describing individual risks or risks that pertain to a specific discipline. For example, it is a mistake to highlight that X millions of dollars of threats and mitigation actions are associated with legal issues. Members of the legal community might take umbrage with having their laundry displayed in public, so to speak.

Distribute Risks to Disciplines

At the RIW, risks were considered in high-level categories. This is for at least two main reasons. First, there typically are too many specific disciplines pertinent to the project to allow each of the areas of expertise hold their own risk reviews. Second, a RIW generates much better results if nondiscipline experts contribute to the conversation. Broad high-level categories engender much more rich conversations than do discussions of very narrowly defined subjects by only the subject-matter experts. So, at the conclusion of the RIW, the RR will be populated by risks identified by the high-level categories.

However, no specific discipline might be identified by the high-level category. For example, if one of the high-level groups is termed "Technology Issues," this could include risks from chemical engineering, research & development, construction technology, and so on. It is the job of the RPP and PM to now subdivide the risks into discipline-specific RRs and to distribute those registers to the discipline manager.

Typically, deciding which risk belongs to what discipline is done not by inspecting the risk itself, but by reviewing the risk action—the mitigation of the threat or capture of the opportunity—and the name recorded as the responsible party (see, it is not just "busy work" when specific risk owners and action takers are discussed and recorded). The names of the people associated with the mitigation/capture action most times will specifically identify to which discipline the risk belongs. This is not always the case. For example, a legal risk might only be resolved by an environmental engineer—listed as the action owner—performing the recorded action. In cases like this, a conversation should be had with both the legal and environmental teams to determine in which RR the risk will be listed. Under no circumstances should any risk be listed twice—that is, in more than one discipline RR.

I have previously in this book briefly reviewed the various types of RRs employed these days (see the "Risk Registers and Their Many Implications" and "Gatekeeper or Free-for-All?" sections of Chapter 2, and the "Lack of

Organization" section of Chapter 5). It is beyond the ken of my capabilities and the scope of this book to address the technical aspects of, for example, Web-based RRs. Just like my car, I use the technology but can't be looked to as a credible source of information regarding how such registers work "behind the scenes."

If spreadsheets are used, then the RPP should keep a "master copy" and distribute subsections of that copy to individual disciplines. A better method, I have found, is to employ the Web-based RR technology. This methodology facilitates the existence of one image of the RR to exist, but allows discipline representatives to view their risks by entering a password (or some other security measure). As discussed in previous sections, the RPP or designated representative should be the only individuals who can update the RR, but by use of passwords and so on individuals can view the risks pertinent to their disciplines.

Individual disciplines should decide which top-ranked risks will be addressed. Those disciplines can also decide to proceed with their own risk-monetization process as described in Chapter 8 of this book.

Implement Sustaining Actions

Following the issuance of results from the RIW as described in the immediately preceding section, it is time to consider—and probably past time—just how the RR will be sustained. That is, how information on existing risks will be updated, how new risks will be identified and added, and other practical matters. All of these topics have been covered in previous sections of this book, but now is the time in the process for implementation. So, I will use just a few paragraphs here to refresh the reader's memory regarding these exceedingly critical steps.

Implement the "Keep It Fresh" Process

In the "Gatekeeper or Free-for-All?" section of Chapter 2, I describe several methods commonly employed to capture new risks. Remember, the RIW is just a snapshot in time of the project. As the project evolves through time, some initially identified risks will be resolved or become moot, but new risks will pop up.

It is absolutely essential that the RPP or designated proxy begin immediately the process of identifying new risks. Regardless of the process selected to collect new risks, the chosen process needs to become part of the fabric of how disciplines work. If specific disciplines are to successfully implement the risk process, then they need to be coerced early in the risk regimen to participate in the methodology to identify and properly record new risks as those risks present themselves through time.

Collecting Information to Modify Existing Risk Information

In addition to gathering data on newly existing risks, the RPP has to implement the selected method of gathering "progress data" on existing risks. In the "Who's the Boss?" section of Chapter 5, and in the "Control of Data Entry" and "New Sheriff in Town" sections of this chapter, I address the issue of how risk information might be updated.

It is outlined in the aforementioned sections that individuals from specific disciplines should not be allowed to update RR information. For example, if a specific risk action is to be completed by a certain date, only the RPP should have the decision-making power to decide whether or not the action was successfully and fully implemented. Make sure that the information-updating process is in place and understood by the organization.

Sending Out Alerts

It is critical that risk-action owners be alerted, periodically, to the fact that the due date for completion of that action is approaching and that the action is not yet complete. Methods for issuing alerts are discussed in the "How People Are Alerted" section of Chapter 5.

It is almost an impossible task for any RPP or group of designates to regularly scour the RR for due dates and action items. If a Web-based RR is implemented, then automatic alerts—such as e-mails—can be sent to responsible parties.

Disseminate Information on Progress

Although the RPP, or anyone else, should avoid making public proclamations regarding the progress on any specific risk, it is the responsibility of the RPP to regularly update the organization regarding progress. Such updates can be in the form of a summary of NPV addressed by threat-mitigation or opportunity-capture actions, capital employed to address risk actions, the number of risk actions completed, the number of new risks identified, and so on.

Such updates should be specifically formatted so that the organization gets used to reading and digesting the information. The updates should go out to all RIW participants and their management as well as to the top management of the project.

Well, Wouldn't It Be Nice?

In this chapter, I have addressed in some detail the processes you might employ if you are lucky enough, as an RPP, to be brought in at the inception of a project. Well, wouldn't it be nice if that was typically the case? Sometimes this happens.

However, it is nearly just as common to be called to service when a project is already underway. Sometimes the project is in quite an advanced state—a state in which project management has finally tumbled to the fact that risks are not being properly addressed and something has to be done. This belated call to duty is the subject of the next chapter.

Suggested Readings

In addition to books and articles cited in the preceding text, this list includes other material that the reader might find helpful and relevant to the subject matter discussed in this section of the book.

Berger, L. A., and M. J. Sikora. *The Change Management Handbook: A Road Map to Corporate Transformation.* New York: McGraw-Hill, 1993.

Block, P. *The Empowered Manager: Positive Political Skills at Work.* San Francisco, CA: Jossey-Bass, 1990.

Koller, G. R. *Risk Modeling for Determining Value and Decision Making.* Boca Raton, FL: Chapman & Hall/CRC Press, 2000.

Koller, G. R. *Modern Corporate Risk Management: A Blueprint for Positive Change and Effectiveness.* Fort Lauderdale, FL: J. Ross Publishing, 2007.

Mourier, P., and M. R. Smith. *Conquering Organizational Change: How to Succeed Where Most Companies Fail.* Newtown Square, PA: Project Management Institute, 2001.

Pasmore, W. A. *Creating Strategic Change: Designing the Flexible, High-Performing Organization.* New York: John Wiley & Sons, 1994.

Pritchard, C. L. *Risk Management: Concepts and Guidance.* Arlington, VA: ESI International, 2001.

7

Another Route to Risk Monetization— The "Already in Progress" or "Imperfect World of Low Control" Path

---

In the preceding chapter, I described a perfect world in which the Risk Process Proponent (RPP) is part of the project team from inception. However, almost as often as the perfect-world scenario occurs, a project team realizes it needs to implement a risk process or is forced to accept implement of such a process. Most of the time, this realization or enforcement is the product of getting into trouble. That is, some significant opportunity is missed ("Why didn't you see that?") or a threat is posed or realized with which they have to contend.

Sometimes, these project teams have done nothing in the way of risk identification, assessment, management, or monetization. If this is the case, then we are back to the perfect-world scenario described in Chapter 6 because even though, as the risk expert, you are not catching the project at inception, you have the opportunity to act as though you have. A far more insidious situation is realized when each discipline has carried out some sort of "risk assessment," but there has been no attempt to bring uniformity to the unique approaches. Now, as the risk expert, you are charged with building a risk process with a plethora of pieces that don't fit together. So, what do you do? I'll explore this hideous situation in this chapter.

Some of the "Perfect World" Stuff Applies

In Chapter 6, nearly all of the advice and processes I describe in some of the major sections still apply. Those major sections are listed below:

- Try to be unfashionably early.
- Appoint a RPP.
- Have "the talk."
- Don't underestimate the upheaval.
- Decide how sensitive "soft" risks will be handled.
- Have the conversation about risk acceptance.

- Determine what has already been done and establish rapport.
- Establish a high-level story about why this is being done.
- Design a risk monetization process that fits with the existing culture and workflows.
- Establish that there will be only "one" risk register.

Although for the RPP it is still essential that he or she comprehend the political, organizational, and cultural aspects of the project team, establish himself or herself as a figure of authority, and other critical aspects delineated in the list of Chapter 6 major sections above, it is highly unlikely that the RPP will be able to persuade project management and discipline experts to submit to the rigorous task of preparing for and attending a Risk Identification Workshop (RIW). The major sections of Chapter 6 that deal with the workshop have, therefore, been eliminated from the list above.

In the situation of risk analyses of various types having already been done—the focus of this chapter—there are at least two primary reasons why a RPP will likely have little luck in conducting a RIW. One obvious rationale used by discipline experts and their management to avoid taking part in an RIW is to persist in the argument that "That ship has sailed." That is, various disciplines (i.e., chemical engineering, construction, law, logistics, human resources, health and safety, environmental, supply chain, research and development, and others) already went through some sort of risk identification process—probably as individual entities. They are not interested in devoting the required time and effort to repeat the process.

While it might be true that some variant of a risk identification method was uniquely employed by the individual cadres of discipline representatives, there likely was no effort to apply a common methodology, or, to rank the risks identified by one discipline with those of another. So, even though the project might benefit greatly from implementation of an RIW as described in Chapter 6, it is not likely to happen.

A second source of resistance regarding execution of a formal risk identification process is the obvious reason—because it takes time and effort. Unlike the perfect-world scenario described in Chapter 6 in which the risk process—right from the start—is part of how the project team does business, the team that is well into the project will absolutely view as an imposition the time and trouble it would take to, in their minds, do something that they already have done. By the time the risk process is called for, many of the team members are quite busy and exceedingly stressed. It is a real challenge to establish a risk process—which will take their time and effort—on top of their already existing litany of tasks.

Purposely, I have painted a bleak landscape in the preceding paragraphs. This is to contrast the focus of this chapter with the previous one. It should be noted, however, that if an RPP can, in fact, persuade the organization to implement the perfect-world scenario outlined in Chapter 6, then that is what he or she should do. This chapter will focus on the situation in which the perfect-world scenario is not attainable.

Risk Register?

Especially in Chapters 2 and 6, I detailed the attributes of the risk register. Therefore, a discourse on that subject will not be presented here. However, the risk register plays just as important a role in the imperfect-world circumstance as in the perfect-world situation.

For project entities that have not built risk registers, then the RIW and risk-register building or populating advice offered in Chapter 6 should be pursued with these groups. However, it is more often the case that each discipline has created some sort of list of risks—usually uniquely threats.

For groups that have generated a list of risks of some ilk, it is the RPP's job to ascertain whether or not the list is up to date and whether opportunities were considered in the creation of the inventory. I truly wish I could offer some really great time-saving advice here, but I'm afraid that there is no easy or quick remedy.

Low-tech as it might be, the only practical way to attempt to assure that the list of risks is current and that opportunities have been considered is to review the catalog of risks with a limited number of representatives from the register-creating discipline. Such a review will have to be performed on a discipline-by-discipline basis. Personally, I have done this many times. There are several salient steps to be taken.

First, the RPP should simply query the group with regard to new risks that might have arisen since the generation of the risk register. This should be a RPP-facilitated conversation in which the RPP collects any new risks in the proper Reason–Event–Impact format as described in the "Risk Description— How to Express a Risk" section of Chapter 2 and exemplified in the "Prepare for Risk Expression" section of Chapter 6. Given that it has to be done sometime, risks already described in the register might be now restated in the correct format. Believe me, this takes time and patience.

Restating the risks in proper format should not be attempted by the RPP alone. I have tried this, and given the curt descriptions often offered in the original risk register, it is impossible for the RPP in isolation to surmise what is the risk event, much less the cause and consequence. However, if other critical fields in the risk register are to be created and populated with information, the statement of risks in the proper format is essential.

Relatively rare is the situation when the original risk register contains many, if any, opportunities, and it is a notable change of mindset for most groups to think of risks as such. Opportunities, however, can have as significant an impact on the value and chance of success of a project as do threats. It is the RPP's responsibility to make example of a few opportunities—even if the examples offered don't directly relate to the current project—and to prompt the group to identify and properly describe as many opportunities as is practical.

If the discipline-specific risk register does not provide for the capture of probability, impact, capture or mitigation actions, action owners, action implementers, action costs, and action due dates, then space for those items must be created in the register and populated with the proper information and data. These attributes are described in detail in various sections of Chapter 2 and exemplified in the "Quantitative Approach" section of Chapter 6. I fully realize that what I am suggesting here—especially on a discipline-by-discipline basis—is a lot of work. This is the price of late entry.

Avoid the "One, Two, Three—Go!" Situation

OK, I am going to assume that from this point on, the RPP has reviewed all risk registers and has brought them up to snuff, so to speak. I realize that to get to the up-to-date point is a real grind. I truly wish I had some sage words of wisdom or some silver bullet to pass along that would alleviate the tedious nature of the effort, but having experienced this situation many times myself, I can assure the reader that time, patience, and effort are the required elements. As I said earlier in this chapter, if you are lucky enough to be able to convince the entire organization that they should follow the perfect-world precepts outlined in Chapter 6—including the holistic RIW— then that route should be taken. However, when groups are already busy and frantic with execution of the project and sometimes believe that they have already addressed the risk issue, it is a tough row to hoe to get them to agree to a perfect-world implementation.

In spite of the great amount of time and effort, the RPP and discipline representatives have invested to get to this point (all risk registers updated and conforming to required standards), little impact on the project can be expected relative to the initial state of the project group prior to updating of the risk registers. This statement might lead you to believe that the update process is a waste of time, but that is not so. Updating of individual risk registers is but a middle step in the process that will result in significant improvement of value and success.

In the "One, Two, Three—Go!" section of Chapter 5, I describe the issue of the relative importance of risks. The perfect-world generation of a single risk register facilitates the relative ranking of all risks—regardless of discipline. Using the one of the ranking processes described in Chapter 6 (see the "Assessing, Valuing, and ranking of Risks prior to Monetization" section), the top N ranked risks will represent those threats and opportunities that, again, regardless of discipline of origin, will have the greatest impact on project value and/or success. Those top-ranked risks are those that will be taken forward to the monetization process and are likely the set of risks that will be addressed with mitigation or capture actions.

Having now updated all of the individual risk registers, the project is now in the situation I describe in the "One, Two, Three—Go!" section of Chapter 5. The RPP is now faced with N risk registers each of which will have some top-ranked risks. What the project team does not want to do is to address the top-ranked risks from each register. This is because the top three risks, for example, from the Environmental Group's register might not, in a discipline-melded register (as in the perfect-world risk register), even come close to being in the top N risks. Addressing risks that will have negligible impact on the value or success of the project—regardless of the fact that those risks were at the top of a specific-discipline ranking of risks—is a horrendous waste of resources and time.

In the perfect-world risk process, I advocated the use of a translation table to use a common metric (such as net present value or NPV) to express the significance of each risk. See the section "Translation of Risks" in Chapter 5 and the "Create Translation Table" section of Chapter 6. I do not recommend that the RPP generate a translation table to address each of the individual-discipline risk registers. There are several reasons for this.

One rationale for avoiding the use of a translation table on individual-discipline risk registers is that the methodologies used to roughly rank the risks and assign value are likely not uniform. In spite of the efforts by the RPP and individual-discipline representatives to apply the risk-register-creating-and-populating best practices to already existing risk registers, the likelihood of that uniformity bridging multiple risk registers is low. So, even though the RPP and others might do their best to have each risk register conform to standards, the dissimilar initial nature of the registers (and the data contained in them) likely cannot be corrected to the extent necessary to apply a single translation method to all registers and have the results make any sense.

A second reason to eschew the application of a translation table to individual risk registers is the fact that some risks will be repeated in multiple registers. For example, the Environmental group might have identified a threat that relates to the increase in costs of construction of waste-processing facilities due to slow government approval of necessary permits. The Construction group might have listed a threat that due to the same government delays that also addresses increased construction costs—including the environmental waste-processing facility. Other groups might also have in some way taken account of the construction costs due to government delay.

I think that it should be obvious to even the most remedial of readers that using a translation table to convert these risks—essentially the same risk—to a common metric for relative ranking is folly. Well, if it is not recommended that the individual risk registers be addressed with a translation table, then what is to be done? The answer to that query is the subject of the next section.

Blending Multiple Risk Registers

If the reader believes that at the beginning of this chapter I offered little hope of being able to convince already-in-action project teams to submit to a new and comprehensive RIW process, just wait until you see the gloom and doom I'm about to relate. There is just no easy way out of this situation.

So, what is the situation? The RPP and discipline representatives have worked—to the extent possible—each individual-discipline risk register into a state of acceptability. Now the project is faced with N (one for each discipline) risk registers that very likely include multiple representations of the same risk (see immediately preceding section). How does the RPP bring these registers together? In my experience, there is only one practical solution, and the reader is probably is not going to like it.

Experience has shown that the most practical method of melding multiple risk registers is to gather together in a room one representative from each discipline and to begin reviewing—one at a time—the risks in each register. This is a low-tech and time-consuming (not to mention mind numbing) process. The aim of such a review is to have, for example, the Environmental Group representative recognize the expression of a risk that is in her or his risk register when she or he sees that same risk described in the register of another discipline.

As the RPP and discipline representatives work their way through the individual risk registers, duplicate risks should be culled and those remaining unique risks should constitute the single, new, comprehensive risk register. Now the question becomes, how can these risks be ranked relative to one another?

During the one-by-one review of risks just described, the RPP and discipline representatives should review what is meant by the term "probability" and what is meant by the term "impact." See the "Review What Probability Means" and the "Review What Impact Means" sections of Chapter 6 for a description of this process. It is essential that this review of risks eliminate, to the extent possible, any differences that might have existed in the estimation of probability and impact when the individual disciplines created their risk registers.

When uniformity of expression of probability and impact in the new risk register has been achieved—to the extent possible—then a translation table should be created and applied in order to express the impact of each risk by a uniform metric (such as NPV). Again, see the section "Translation of Risks" in Chapter 5 and the "Create Translation Table" section of Chapter 6 for explanations and example of how to create and apply a translation table. The common expression of impact (such as NPV) can be used to roughly rank risks and to create a list of top-ranked risks that will be carried through to the risk-monetization process. That process will be the subject of the next chapter.

"Risk Analyses" Already Done

Thus far in this chapter, I have described a situation in which the realization for the need to apply a risk process comes to a project group before it is

too late to consider creating new and/or fixing already existing risk registers. This represents the imperfect-world best-case scenario—even though it
might not seem like such a pleasant situation.

Worse, from the RPP and overall-project-benefit point of view, is the situation in which individual disciplines have already ranked their risks and
have expressed to management—again, individually—the impact of the
risks on the project. Actions to mitigate threats and to capture opportunities
might also be in progress or completed. The point at which the RPP might be
called in to implement a risk process can, of course, be anywhere in project
life from inception to near completion—a broad spectrum of entry points.
It is beyond the scope of this book to address the myriad opportunities for
intervention. I will describe here, what seems to me, to be the last point at
which a RPP should consider helping a project group before that RPP refuses
to get involved because it is effectively too late. This is, then, essentially the
opposite end of the spectrum relative to being involved in the project from
inception.

Individual Expressions

Left to their own devices, individual discipline groups usually will express
the probability and impact of risks in ways with which they are most comfortable. For example, the legal team probably will use prose (text) to relate
the probability and impact of legal risks and is unlikely to generate probabilistic cumulative frequency plots from a Monte Carlo process. However,
such plots are exactly what an engineering group might use to present their
risks to management. These two methods likely represent the two ends of
the sophistication-of-representation spectrum regarding risk presentations,
but they are not uncommon.

In Figure 7.1—which is Figure 1.2 repeated here to save the reader from
having to page back to Chapter 1—I illustrate just a few of the more popular methods for expressing the probability and impact associated with risks.
Working under the assumption that addressing and melding risk registers is
a ship that has sailed, the RPP can find herself or himself in the situation of
attempting to integrate such expressions of risk.

Other expressions of risk also are common such as the many variations on
the "tornado" diagram. Such representations are, however, aimed at a specific
aspect of the risk. For example, the tornado diagram is an attempt to reveal
the "sensitivity" of the output parameter (the "answer" variable) to any individual input parameter. These sorts of "specialty plots" will not be directly
addressed here, although they have their place in the scheme of things.
See the "Sensitivity" section of Chapter 5 (Figure 5.5) and the "Sensitivity
Analysis" section of my book: *Risk Assessment and Decision Making in Business
and Industry: A Practical Guide*—2nd Edition (see References at the end of
this chapter) for details regarding the role of tornado diagrams and other
methods.

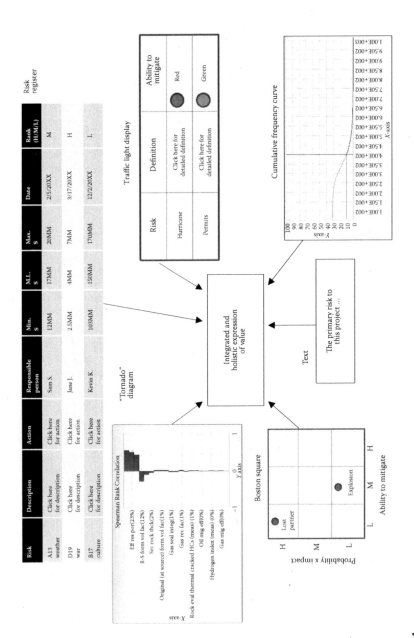

FIGURE 7.1
Example of the expressions of risk that must be integrated to get a holistic view of perceived value for a project.

Really, meaningful integration of such representations of risk—as shown in Figure 7.1—is a highly improbable if not impossible task without resorting to tracing the lineage of each expression back to the risks—at least probabilities and impacts—that created the expression. If that can be done—effectively, going back to the risk registers—then the steps outlined earlier in this chapter should be applied. In order to represent the far end of the spectrum, however, I am going to assume that retracing the plots back to the risk registers is not practical for technical reasons and/or lack of time.

Before you believe that I am giving up without a fight, I would point out that sometimes the expressions of risk—especially tornado diagrams or cumulative frequency plots—are the result of complex integration of probabilities and impacts from multiple risks using hundreds and sometimes thousands of lines of computer code. So, the final expression of the complexly integrated risks—such as a cumulative frequency plot resulting from a convolute Monte Carlo program—can't practically be broken down into the contribution of the individual risks. If the RPP does not have access to the input data for such computer programs, or, there simply is not time or resources to retrieve and process the original data, then the RPP has to do what she or he can with the result.

In the following sections, I offer suggestions regarding conversion and integration of just a few of the more common expressions of risk. These suggestions should not be taken as doctrine but should be viewed only as examples of what might need to be done depending upon the project and how the risks were integrated.

Converting the Cumulative Frequency Plot

In the "Chance of Failure" section of Chapter 3, I address the types of cumulative frequency plots that can result from considering or not considering chance of failure. The two plots in Figure 3.7 are representative. In this example, I will ignore the impact of integrated individual-parameter chance-of-failure values on the Y-axis intercept of the cumulative frequency curve (and of other effects of chance of failure), but the reader should keep in mind that the issues associated with chance of failure add another layer of obfuscation and complexity to the discussion to follow.

A cumulative frequency curve such as that shown in Figure 7.1 might result from a Monte Carlo process such as that described for the top plot in Figure 3.7. The reader is referred to that section of Chapter 3 to review how such a plot might be produced.

It might be decided that NPV is the common metric to be used to express an individual-risk impact, or, the impact of a combination of risks. If the X-axis of the specific-discipline cumulative frequency plot in Figure 7.1 was NPV, then as an RPP, I'd consider my job done. However, let's assume that the X-axis is not NPV but represents the probable costs to the project if the N risks integrated by the Monte Carlo program were to materialize (this is not the same as the cost to mitigate the risks).

What we have, then, is a range of probable cost associated with all of the risks that were addressed in the Monte Carlo program. The individual-risk probabilities and impacts—likely both represented as distributions—were integrated by the Monte Carlo code to produce the plot. So, if we want to use NPV as the metric to express the impact of this individual discipline's integrated risks (remember, we are operating under the assumption that we can't practically go back to the original data), how can we get from this plot of costs to one of NPV?

There exist a wide range of mostly convolute routes to the end. I will here describe the most straightforward and simple approach.

Every cumulative frequency plot can be expressed in frequency space (i.e., a "distribution"). In Figure 7.2, I show a cumulative frequency plot and its frequency equivalent. In addition, every project has associated with it an economic model. Typically, the model is kept by the Commercial Analyst or some other individual or group that was originally charged with calculating the NPV (or whatever financial measure is to be used such as IRR or DROI or whatever—we will assume here NPV) for the project. This model usually, but not always, is a giant spreadsheet or complex linkage of multiple spreadsheets.

Any spreadsheet can be linked to a commercial software package—and there are several very popular spreadsheet add-on applications—that can use the frequency plot shown in Figure 7.2 as an input distribution. The add-on packages facilitate the sampling of the distribution N times (typically, 1,000 or more "iterations"—a value set by the user). Each time the distribution is sampled, the value selected is input to a cell (or distributed over a series of time periods) in the spreadsheet and, in this example, the resulting NPV is calculated. The probability of any given revenue is handled by the concentration of revenues on the frequency-plot's X-axis (which, along with "bar" or "bin" width, controls the heights of the bars in the frequency

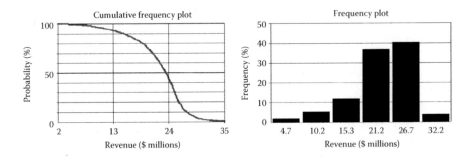

FIGURE 7.2
A cumulative frequency plot and its frequency equivalent.

plot). Each NPV value from this process is saved and a cumulative frequency plot of NPVs can result. This plot, then, can be integrated later with similar expressions of NPV from other disciplines. I'll explain that process later.

Converting the "Boston Square"

For the reasons that I detail in the "Use of the Boston Square" section of Chapter 5, I am dead set against the use of the Boston Square. I would rarely advocate its use, and the fact that I am writing about it here is a bow to reality but should not be interpreted as endorsement.

I am enough of a realist to know that in spite of my well-founded misgivings regarding the Boston Square, it is in widespread use. The reader is likely to encounter the misguided plot. It is my responsibility to offer recommendations about how to convert such plots into some measure of risk, and therefore, project value.

Admittedly, there is at least one beneficial aspect of the Boston Square—at least relative to the cumulative frequency plot discussed in the previous section. When viewing the cumulative frequency plot, it is impossible, by inspection only, to identify any specific risk or the impact of any single risk or group of risks. When inspecting the Boston Square, each symbol (usually a dot or square) on the plot typically represents an individual risk. This eliminates the need to try to "back out" of a risk expression (like the cumulative frequency curve) the attributes of any given threat or opportunity.

A downside to the Boston Square is that the Y-axis typically represents the product of probability and impact. For the purposes of valuing and ranking risks, the X-axis—usually an expression of the project team's ability to mitigate the threat or capture the opportunity ("Manageability" is a common X-axis label) is superfluous to our effort to value and rank the risks.

Y-axis values on a Boston Square represent probability-weighted impacts. So, if a risk represents a "whopping big" impact but has associated with it a "tiny" probability (low-probability or high-impact risk), that risk would plot near the mid point (or so) on the Y-axis. The Boston Square is, essentially, a 3-D plot in 2-D.

Given that someone plotted the risk on the Boston Square and given that there are no hair-raising sets of equations to be navigated in order to plot a point along the Y-axis (just probability × impact), it generally is not too onerous a task to discover the probability and impact values that were combined to generate a Y-axis value for any plotted risk.

When the individual-risk probability and impact values are determined, these values can be used in several ways as input to the same project economic model (that produces NPV numbers as output) as described in the section "Converting the Cumulative Frequency Plot" of this chapter.

Using one of the spreadsheet add-on programs that can cause a spreadsheet to be part of a Monte Carlo process, the probabilities and impacts can be used as deterministic values. That is, on each iteration of the Monte Carlo process, by comparison of the risk's probability value against a random number in the range of 0–1 (or 0–100), it can be determined whether or not the impact of the risk—let us say a cost—will be considered in the calculations of that iteration. If the probability test determines that the cost should be considered, then the cost coefficient is inserted into the appropriate cell (or cells in the case of a time series) in the spreadsheet and the resulting NPV will reflect the risk's impact. The reader should be sure that if the cost, in this example, will not be experienced in a single time period in the spreadsheet model, that the cost is appropriately distributed across the correct time periods. It is beyond the ken of this book to delve into the programming details of how this might work for any given spreadsheet, but keep in mind that an NPV is the result of processing discounted cash flows from a number of time periods.

It breaks my heart to be recommending the use of spreadsheet add-on programs to cause the spreadsheet to take part in a Monte Carlo process. I know that those programs do not handle—or have an exceedingly tough time handling—things like chance of failure, resampling a single distribution multiple times within a single iteration (one sample from the distribution for each time period, for example) and fall short in a number of other ways. However, again being a realist, I know that it is with these sorts of programs that the reader will likely have to contend.

One upside to such add-ons is that they typically facilitate the expression of any parameter as a distribution. Being a "probabilistic purist," I would advocate that when the probability and impact single values are discovered from each risk's Boston Square Y-axis value, those values be expanded into ranges or distributions. This can be done simply through conversation and the process for collecting range data is described in the "Probability of Occurrence" section of Chapter 2. Using ranges instead of deterministically described probabilities and impacts in the Monte Carlo process described above can give insights into the uncertainty associated with the risks. In the examples given in this book in which I have selected NPV as the common metric, those NPV values can be used as the risk-ranking tool. More on this, later in the "Risk Monetization" chapter.

Converting the Traffic Light

Traffic light processes are ubiquitous when it comes to "ranking" risks. I put the word "ranking" in quotes here because the typical traffic light has only three colors, and therefore, just three categories by which risks can be ranked. Such ranking is crude, but sometimes all that is required.

Whether or not three categories is sufficient to represent the range of probabilities and impacts of risks, if the task is to rank traffic-light-represented risks with other differently represented risks, then, just like the cumulative frequency curve and the Boston Square addressed previously, this three-colored description of a risk will have to be converted to a common metric. In all of the examples I employ, I am using NPV as that common measure.

Consider the traffic light representation in Figure 7.1. I realize that these books are printed in black and white, so talking about what is the color of the light in the figure is a bit silly. Let us assume that the color of the light associated with the Environmental Risk in Figure 7.1 is red.

Red typically is the color used to express "high risk." What the users really mean by this is either that the threat or opportunity has associated with it a high probability, a high impact, or both. More often than not, the light is an expression of impact, but you cannot count on that. Any RPP attempting to convert a traffic light to a common metric to be used for relative risk ranking should make sure the intent of use of the colors.

Use of the color red to represent the environmental risk means that this risk is something to which the project team should pay attention and about which they should do something. I know that I don't have to point out to anyone that if you enter "red" into a cell in the typical spreadsheet economic model, the calculations will not go so well. So, how can a RPP convert this color to something that can be used in the spreadsheet equations?

First, the RPP needs to have a conversation with the person or group responsible for assigning the risk a red color. If the risk represents an environmental threat, for example, that will impact the project over time; the RPP might come armed with the same three questions for each time-related and traffic-light-represented risk. These queries might be as follows:

- What is the probability that this risk will occur?
- If the risk does materialize, for how long will we be affected by the impact of the risk?
- How many dollars per day (or week or month or whatever are the time increments) is this likely to cost us?

Getting answers to these questions in the form of deterministic values or ranges will allow the RPP to input the values into either a simple equation such as:

Impact of risk = (Length of impact time) × (Cost of impact per unit time)

This, combined with the probability value can be used to get rough estimates of the risks impact (NPV) on the project when the coefficents are used as input to the aforementioned and described Monte Carlo process facilitated by linking the spreadsheet add-on program to the project economic model. Although I advocate the use of a Monte Carlo model—as is my wont—keep in mind that single values resulting from the questions above also can be input to a spreadsheet economic model to produce a single NPV. Also remember that if the "Impact of Risk" shown in the above simple equation is time dependent, then spreading that impact over the appropriate time periods in the model will be necessary. More on that in the next chapter on risk monetization.

In this example, I have addressed just one type of risk—one that has a cost as an impact and one that has an associated time element. A slightly different trio of questions will have to be devised for other types of risks. The spectrum of risk types is broad. It is beyond the capability of this author to even come close—in the limited space and time allotted by this book—to consider the specific questions that might have to be devised to fully address all of the risk types. It is my hope that the reader who might find herself or himself in the position of a RPP will be able to grasp the concept of generations of queries for traffic light conversion and use that concept to create questions appropriate for the project at hand.

Converting Text

I know that it might seem that I have saved the easiest conversion example for last. However, it has been my experience that conversion of text to, for example, NPV can be the most vexing of translations.

First, I am always thinking: "What the heck did they mean when they said that?" Use of text to describe risks is a favorite ploy of legal departments. Not only, then, can the text be long and convolute, but the jargon utilized supposes some knowledge of the domain. In Figure 7.1, the partial textual description of the risk is as follows:

> "In our opinion, the primary threat to timely construction of the plant
> is the question of ownership of land and transport capability combined
> with the unsettled question regarding land-based access ..."

This type of risk description can go on and on and generally combines expression of more than one risk. I don't think there is any nefarious intent in combining multiple risks in a long statement or series of statements—it is just the nature of the beast, so to speak. Intentional or not, it is the job of the RPP to attempt to untangle this web of intrigue.

I wish I had some magic bullet to offer here, but there is just no getting around it. To convert such text to a common metric for use in risk ranking, the general steps to be taken are the following:

- Gather together the individuals responsible for penning the text.
- Through conversation, determine the number of specific risks that are represented in the text and record those risks in the proper Reason–Event–Impact format.
- Again, through conversation, glean for each risk a deterministic values or ranges that represent the probability and impact of each risk.

- Just like in the traffic light example above, insert these single values or ranges into the economic spreadsheet model in order to generate a rough estimate of the risk's NPV impact.

Post Conversion

As I stated at the outset of this major section, conversions of the type described above are a last-ditch and rather desperate attempt to bring uniformity of expression to a field of risk expressions that are disparate and ignorant of one anther. In spite of such conversions to a common metric, the meaning of the value of any group of risks can be "all over the map" because the original expression—in the cases above, the cumulative frequency plot, the Boston Square, the traffic light, or the text—is difficult to completely eradicate.

Carrying these risks through to a monetization process, as described in the next chapter, will not improve the original-expression-eradication problem. However, as will be revealed, a more accurate estimate of the impact of any single risk or set of risks can be had if the monetization process is employed.

Suggested Readings

In addition to books and articles cited in the preceding text, this list includes other material that the reader might find helpful and relevant to the subject matter discussed in this section of the book.

Berger, L. A., and M. J. Sikora. *The Change Management Handbook: A Road Map to Corporate Transformation*. New York: McGraw-Hill, 1993.

Jasanoff, S. "Bridging the Two Cultures of Risk Analysis." *Risk Analysis 2* (1993): 123–129.

Koller, G. R. *Risk Assessment and Decision Making in Business and Industry: A Practical Guide*. 2nd Ed. Boca Raton, FL: Chapman & Hall/CRC Press, 2005.

McNamee, D. *Business Risk Assessment*. Altamonte Springs, FL: Institute of Internal Auditors, 1998.

Williams, R. B. *More Than 50 Ways to Build Team Consensus*. Andover, MA: Skylight Publishing, 1993.

8

Risk Monetization

Well, if you are atypical and have read this book from the beginning to get to this place, you might find this chapter rather anticlimactic. While it is true that the risk-monetization process is a key to project-execution efficiency, just like painting the inside of your house, it is mostly about preparation.

A Warning and Justification for Monetization

All of the chapters leading up to this one—especially Chapter 6—describe the tedious, complex, and mentally and physically exhausting task of preparing for risk monetization. Again, just like painting, after all of the furniture moving, nail pulling, crack patching, sanding, cloaking of floors, clean up, and so on, the painting step—although the essential aspect and the primary aim of the entire effort—seems rather mundane in comparison with the combined preceding efforts. So, it will seem to the reader regarding risk monetization relative to the efforts required to prepare to monetize risks.

Correlated Risks

I will present here just a warning about a considerable problem that impacts not only the risk-monetization process, but any process that attempts to combine parameters that might have a "relationship"—that is, that are not independent. To the uninitiated, this subject might seem somewhat trivial, but I can assure the reader that personal experience has shown this issue to be of great significance.

When most people in the risk business say the word "correlation" what many of them really mean is that there exists a relationship between two (or more) parameters and they would like to honor that relationship in any set of calculations (i.e., model). That is, if the parameters are represented as ranges (distributions) in the model, they would like the value of one parameter to "track" or make sense with the chosen value for another parameter.

For example, it might be that two parameters in a model have a relationship. These parameters might be Depth and Temperature (see Figure 5.5 in Chapter 5). Depth is a measure of how far below the surface of the earth we are taking temperature measurements. Temperature is the measure of, say, the degrees Fahrenheit at a given depth. Our depth scale might go from 0 to 20,000 ft. The associated temperature scale might range from 60°F at the surface (0 feet of depth) to 500°F at the terminal depth of 20,000 ft.

What one is hoping to avoid, in selection of temperature and depth values to be put simultaneously into an equation, is the selection of a depth–temperature pair that does not reflect reality. For example, one would like to avoid selection of a depth of 19,000 ft and a temperature of 70°F. What is desired are depth–temperature pairs that correspond to the relationship that these parameters exhibit in nature.

Correlation is just one way to exhibit some expression of the "tracking" of one parameter by another. A correlation coefficient is nothing more, in practice, than an indication of how good a predictor of the value of a second parameter is a known value of a first parameter. For example, if I know the depth value in the example given above, how good a predictor of temperature is that known depth value?

A relationship between two parameters might not be linear or easily calculated. Parameter relationships in the field of psychology, just for example, are real, but often difficult to predict with even the most sophisticated models. It will be assumed for this chapter, then, that any relationships between parameters have to be dealt with in the computer code of the model. In addition, it is assumed that the risk-process proponent (RPP) or anyone else has the wherewithal to recognize when risks might be "correlated"—that is, have a relationship—and insist that the appropriate code be generated to honor that sometimes quite complex connection. In this chapter, it will be assumed that this is being accomplished and the examples given here will address independence of parameters. See my book *Risk Assessment and Decision Making in Business and Industry—A Practical Guide*, 2nd Edition (referenced at the end of this chapter) for a more full treatment of correlation.

Rough Estimates of Risk Impact Often Misleading

Especially in Chapters 6 and 7, I delineate methods of roughly ranking risks by a common metric. Processes followed in the "perfect-world" scenario of Chapter 6 might give more realistic estimates of risk impact than those described in the "imperfect world" situation addressed in Chapter 7, but all methods of impact-on-project-value described thus far will yield only rough guesses at risk impact.

Given the aim of evaluating risks thus far—to rank them relative to one another—the rough estimates are fine. For example, I might pick up a standard pencil and declare that it is "one foot in length." Of course it is not. However, I proceed to use the same pencil to measure the height of each person in the room with the aim of lining these people up in height order.

So, I measure the height of the first person with my "foot-long" pencil and that person turns out to be 13 "feet" tall. All other people are measured in the same way and similar coefficients for "feet" are generated. Well, I don't know about you, but I don't know too many folks who are 13 ft in height. Clearly, I am using an inaccurate measuring device. However, if I use the same device to measure the "height" of each person, then I can still accomplish my goal of ranking them by height. So, consistency, in this case, trumps accuracy.

So it is with the rough measurements of, say, NPV for each risk thus far. The NPV associated with a risk might not be the best estimate of risk-related impact on NPV that we could generate, but if we use the same method to estimate NPV for each risk, we should, just like the mismeasured people, be able to use that NPV to roughly rank the risks relative to one another.

Ranking is an absolutely critical element of the risk-management or risk-monetization process. As described in the section "One, Two, Three—Go!" of Chapter 5 and the section "Avoid the 'One, Two, Three—Go!' Situation" of Chapter 7, what most project teams cannot afford to do is address each and every risk. Even risks that rank at the top of a list of risks for a given discipline might not be addressed if they rank low relative to all other project risks. There is only so much money and time to be spent on threat mitigation or opportunity capture and the typical holistic risk register can contain, literally, hundreds of risks. Putting time and resources toward mitigating those threats and capturing those opportunities that will have the greatest impact on project value and/or success is essential. Therefore, ranking is critical even if the consistent measurement of impact of the risk is not as accurate as we might like.

As indicated by the heading of this section, the initial estimates of risk impact can not only be rough, but can be somewhat misleading. In most of the risk-ranking methodologies described in previous chapters, I made scant reference to the time value of money. Accounting for the value of capital relative to time is critical for more accurate estimates of risk impact on the project.

For example, the risk register might include two opportunities that both indicate that if they are captured, they would both yield an additional $100,000 in revenue. Without considering the time element, inserting $100,000 in benefit for each risk into the project economic model will yield the same increase in NPV for both risks—all other things being equal.

Imagine, however, that the economic model is considering 30 time periods with each period being 1 year. The model, then, is said to have a 30-year time series. It is beyond the scope of this book to launch into a meaningful discourse regarding discount rates and the time value of money. However, I will say that because of the use of time periods and discount rates, $100,000 spent all in the first year of the time series will have a significantly greater impact on NPV than will $100,000 spread out over the final five time periods of the 30-year series.

So, even though we might have determined that two opportunities each will yield an additional $100,000 in revenue and used this measure to generate a crude estimate of NPV impact for the risks (by employing the risk's probability and a simple translation table, for example), unless we use more sophisticated means to account for the distribution of those $100,000 estimates over the proper time periods, the estimates utilized to rank the risks will be misleading with regard to absolute impact on the project by the risks. While considering the risk's probability, the simple translation table process utilized is constructed to convert time impacts and various other measurements (tons, utilization, labor costs, etc.) into the common metric without rigorously considering the differences between the various expressions of impact nor the time value of money. Therefore, any decisions other than rough ranking to be made regarding, for example, which risks to address and how much capital it will take to address them requires more rigorous processing—the risk-monetization process.

In addition, in the previous chapters, I have indicated that it injects a complicating factor when impacts are multiplied by probabilities prior to ranking of the risks. One reason for eschewing the practice of generating a product is the low-probability or high-impact issue. As described several times in previous sections of this book, project team members typically want to be cognizant of, for example, threats that could have huge impacts on the project. If those high-impact threats are multiplied by very small probabilities, the resulting product can't be distinguished from products that result from "middle-of-the-road" impacts multiplied by "middle-of-the-road" probabilities.

Also, the decision of whether or not to address a risk is not just a matter of the risk's ranking in a list of risks. The issue of cost–benefit also comes into play. Unless a risk represents the type of issue that absolutely has to be addressed—such as most health & safety problems—then the decision regarding whether or not to mitigate a threat or capture an opportunity will be influenced by how much it will cost—in time and/or resources—to mitigate or capture. For example, if the action item to address a risk is going to cost three times the impact of the risk should it materialize, it is unlikely that the risk will be addressed. If the impact of the risk has been disguised by affecting it with the associated probability, then decisions regarding the cost–benefit and be difficult or impossible to make.

All things considered, the bottom line is that a project team has to make informed decisions regarding which risks to address and how much capital should be spent in the mitigation or capture of each threat or opportunity respectively. Risks that affect health and safety might be addressed (almost) regardless of the consideration of capital. That is, if a risk poses a threat to the lives of workers, then that risk certainly is addressed. If the cost of addressing that risk is deemed too high, then it must be considered whether or not to proceed with the project. Doing something about other risks—even those that do not threaten life and limb—might be mandated by law or other means. Risks not falling into either of these two categories typically are evaluated by their impact on project value or success. The vast majority of risks fall into this later category. It is to those risks that the remainder of this chapter will be dedicated.

Setting the Stage

In previous sections of this book, I have somewhat painstakingly described what I deem to be the most important aspects of preparation for execution of a risk-monetization process. It is no mistake that most of the volume of this book addresses the human, organizational, political, and cultural arenas. As I have stated several times, technical prowess is moot if the organization into which the new technologies or methodologies are to be injected is not properly assessed, understood, and prepared. I have offered advice and examples of how these things might be accomplished and will for the remainder of this chapter assume that such preparation has successfully been executed.

As with most other things in life, illustration by example is a most effective means of communicating convolute processes or ideas. In adherence to this precept, the remainder of this chapter will be presented in the form of an example of application of risk monetization. Because I write these books so that any person in nearly any aspect of industry, academia, or government, hopefully, can translate the lessons offered to their particular situation, the example utilized here will be as generic as is practical to present. Clearly, if there is to be an example, then I have to select a single example project (profound, huh?), but I will take great care to keep the applications of the monetization process within that example as general as can be.

Description of the Risk-Monetization Example

There is not room in this book—or in any book—to make example of every conceivable situation for any domain (government, academia, industry, etc.). Therefore, this is one of those situations in which no matter what example I choose to illustrate the risk-monetization process it will not make most people happy. Given that, I believe, most readers of this book (assuming there are any) will represent industry, the example I will utilize will be drawn from that arena. However, the risk-monetization principles applied will be generic and can be employed by nearly any discipline.

Construction of a new hotel in this country will be used to illustrate the principles and processes of risk monetization. It is assumed in this example that the RPP has executed the steps outlined in Chapters 6 and 7 regarding identification of risks, building a cogent risk register, and using the data captured in that register to relatively rank the risks.

In Figure 8.1 is the top part of a risk register that was created for the project. The four risks shown in Figure 8.1 are just four of the project's many top-ranked risks. In that register, it says, "Click here for details" under the Complete Description heading. For each risk, the complete descriptions

Risk ID	Concise description	Risk description	Probability (%) (COF %)	Impact units	Impact minimum	Impact peak	Impact maximum	Translated impact (NPV in $MM)	Mitigation/ capture action	Probability of action effectiveness (%)	Action cost	Action owner	Action implementer	Due date
ENV 3	Water runoff – cancelled permit	Click here for full description	COF 40						Early talks with local legislature at highest levels	60	$10,000	CEO	CEO	1/1/ 20XX
CON 15	Zoning laws – parking facility	Click here for full description	Prob. 50	$ CAPEX in 1st Yr.	$500,000	$550,000	$580,000	–3.9	Lobby zoning commission	$80	$40,000	Company head negotiator	Project team negotiator	12/10/ 20XX
COM 7	Sports arena construction	Click here for full description	Prob. 60	$/year increase	66,000	200,000	340,000	2	Early talks with city officials	80	$30,000	Commercial analyst	Commercial analyst	2/31/ 20XX
COM 27	Competitor builds hotel	Click here for full description	Prob. 10 COF 18	$/hr. difference/ Year all employees	240,000	480,000	720,000	–0.9	Early submission of plans	90	$120,000	Head planning dept.	Project team planning lead	3/13/ 20XX

FIGURE 8.1

Risk register showing the four of the top-ranked risks for the project. Column labeled "Translated impact (NPV in $MM)" contains values that represent the difference between the base-case NPV ($20 Million) and the mean probability-impacted value generated from the risk's probability and it's impact value as determined from the translation table. The time value of money has not been taken into account.

would be as follows:

- Risk ID ENV 3: Water runoff from the paved hotel property will exceed current acceptable limits. If a variance permit is not granted, then the entire project will be cancelled.

- Risk ID CON 15: If zoning laws are not changed so that a multistory parking facility can be built, then additional already-in-use land will have to be purchased for parking. This will significantly negatively impact the first-year Capital Expenditure (CAPEX) and, as a result, negatively impact NPV.

- Risk ID COM 7: If a proposed sports arena and associated businesses become a reality in the area of the hotel, utilization rates—which are currently projected to be barely adequate—might be positively impacted over some or all of the time periods considered in the project economic model. This could appreciably impact revenues and NPV.

- Risk ID COM 27: If another competitive hotel chain is successful in obtaining large tracts of already-in-use land in the area, then labor rates will become more competitive and will impact revenues and NPV. If, because of the building of the competitor hotel, our labor rates are projected to be prohibitively high, then our project—if in its initial stages—might be cancelled.

In the list of risks above, the risk ID labels relate to the high-level groups at the Risk-Identification Workshop (RIW). In this example,

ENV = Environmental high-level-group risk

CON = Construction high-level-group risk

COM = Commercial high-level-group risk

Note that the data collected in this register emanated from the small cadre of individuals gathered to address a finite number of high-level risk categories. The values shown in the risk register are used to roughly rank risks—relative to one another—within the register. Later, when the RPP has more in-depth discussions with risk-discipline experts regarding the top-ranked risks, data certainly will be changed or updated. The new and updated information will be that used in the actual risk-monetization process described later in this chapter.

Types of Chance of Failure

A summary of the major attributes of chance of failure (COF) is presented in the section "Chance of Failure" of Chapter 3. Details associated with the COF concept will, therefore, not be reiterated here.

COF can emanate from at least three major sources. The three roots of COF briefly described here are in no manner meant to represent a comprehensive cadre. Sources for COF can be as varied as are the types of projects to which COF is applied.

COF that is not associated with a distribution or range is one type. In the example presented here, this variant of COF is represented by Risk ID ENV 3 in the risk register and in the list of risk descriptions in the immediately preceding section.

In effect, this type of COF says, "If this happens, it's over." Associated with this type of COF is only a probability. The probability can be represented as a deterministic value or as a range (distribution). For example, members of the project team might state, "There is a 30% chance that the legislature will not grant the environmental variance required for us to proceed with the hotel." Alternatively, the same committee might relate the probability to a distribution (range) of values that might, for example, range from a low of 20% to a high of 50% with a "most likely" (where is the "hump" or high spot in the distribution) of 30%. As will be made clear in a subsequent section, this type of COF will not be used as a multiplier, but it will be integrated with other COFs to calculate the chance of success (COS) for the project. That COS is expressed as the Y-axis intercept of the cumulative frequency curve. More on this will be discussed later.

A second type of COF is that which is derived from part of an input distribution. The three distributions shown in Figure 3.6 in Chapter 3 (not repeated here) are exemplary. In the porosity distribution shown in Figure 3.6 for example, the entire range of porosities that might be encountered in the real world might range from 1% to 20%. However, experience has shown that if the porosity actually encountered is 5% or less, then the project is a failure. The area of the frequency plot to the left of the cutoff value represents the % COF for that parameter.

In our current example, the RISK ID COM 27 in the list of risks above is an example of a risk associated with the just-described type of COF. In Figure 8.2 is shown the distribution of possible labor costs. The upper end of this range represents labor costs that would be, in management's view, too high to sustain profitability in the long run. Such high labor costs might be forced by a competitive market for labor if a competing chain hotel is successful in obtaining sufficient tracts of already occupied land in the area and begins to build a hotel. The area of the frequency plot to the right of the cutoff (cutoff is the upper end of acceptable hourly-wage values) is as it will turn out, part of the COF associated with labor costs. If our project is in its initial stages when and if the competition successfully obtains property on which to build, then the construction of our hotel will be reconsidered, and, might be cancelled (abject failure of the project).

In the Monte Carlo process, the distribution shown in Figure 8.2 will be used as input. However, random samples for use in the Monte Carlo process will be drawn only from those distribution values less than the cutoff value. The area represented by the bars to the right of the cutoff will be used to calculate the COF for Labor Costs and that COF will be integrated with other

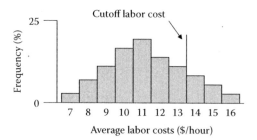

Average labor costs above the cutoff are projected to
cause failure (cancellation) of the hotel project if it is
in its early stages.

FIGURE 8.2
Distribution of average labor costs showing "cutoff" value, which represents project management's acceptable limit.

considerations and with independent COFs (such as the one related to the
environmental-variance COF described above) to calculate the COS for the
project.

In the two examples given above regarding COF, both types of COF
resulted in abject failure of the project. That is, the project was "killed" due
to the COF condition being true. However, there exists another rather ubiquitous type of COF that is a bit more difficult to define even though it is
exceedingly common.

Failure to obtain the environmental-regulation variation in our example
will cause the hotel to not be built. Projected high labor costs resulting from
a competitor hotel being built in the area might cause our hotel to fail to be
built. Both of these types of incidents represent abject failure of the project.
Consider, however, the scenario in which, say, materials need to be delivered
to the construction site by a given date. Delivery of those materials by that
date will be considered failure with regard to the current-year's financial goals
because late delivery will result in significant increases in capital costs, schedule delays, and, ultimately, operating costs. Although, perhaps, not sufficient to
"kill" the project, this type of late delivery, for example, is failure with regard
to short-term goals such as this year's budget and schedule. The hotel-building
example delineated here does not include such a COF, but it is not uncommon
in industry to experience failure to meet quarterly or yearly goals because of
the COF associated with a critical element of the quarterly or yearly effort.

The Monetization Process

Ah, and now the moment you have all been waiting for—a description—
and example of the risk-monetization process. Having in the preceding

sections set the stage for the monetization effort, the following sections will detail how such a method might be carried out using our hotel-building example.

New Conversations

In Chapters 6 and 7, I advocated the use of data-collection and translation methods to gather at the RIW initial and likely crude estimates for coefficients that pertain to the impact and probability related to a risk. Also recommended was the utilization of risk probabilities and a rather crude translation table, for example, to transform initial estimates of impact into a common metric (such as NPV). These relatively rough estimates of probability affected NPV impact are shown in the risk register in Figure 8.1.

The values shown in the column entitled "Translated Impact (NPV in $MM)" represent the difference between the project's deterministic NPV base-case value and the mean of the probabilistically affected NPV-impact value for a risk. It is now time for the RPP to confer with specific-risk experts with the aim of confirming or, sometimes, radically changing the probability and impact coefficients that were captured in the RIW by a group of individuals most of whom were not experts in the area of the considered risk.

This is not at all to indicate that initial estimates of probability or impact were useless. To the contrary, those estimates will form the basis of the conversation between the RPP and the identified discipline expert.

To exemplify these new conversations, let's consider the risk CON 15 in the list of our top-four-ranked risks. This risk relates to the threat of zoning laws not being modified in our favor. If we do not experience a favorable modification to the zoning laws, the project will not be killed, but first-year CAPEX expenditures will significantly increase due to the purchase of parking-lot land to substitute for the multilevel parking facility that is prohibited by existing zoning regulations.

Remember, at the RIW, the high-level group was comprised mainly of individuals who might not have been experts in construction nor in local zoning regulations. The estimates of probability and impact for the construction risk labeled CON 15 might require honing by further conversation with an expert in the area. In addition, a significant portion of time might have passed between the time of the RIW and the risk-monetization effort. Things change. A conversation between the RPP and the discipline expert can bring up to date the estimates of probability and impact.

A conversation between the RPP and discipline expert should be one of in-depth discussion of the risk with the intent to update the information captured in the risk register at the RIW. Sometimes such verbal exchanges require that initial estimates be completely changed. If this rare occurrence presents itself and if the new estimates for probability or impact would cause the risk to be considered less critical to the project, then it is the RPP's job to use the new data to rerank the risk (relative to all other risks) to determine whether the risk is still in the top N risks to be considered in the risk-monetization process.

Conversations between the RPP and discipline expert typically result in new, but not astoundingly different values being used to represent the impact and probability associated with a risk. It is critical, however, that the time element associated with the risk be established.

For example, at the RIW, the impact of a risk might have been determined to be, say, $2,000,000 (two million dollars). Using a risk's probability and the crude translation table to convert this value to NPV, the resulting NPV impact might have been $400,000. The translation table, however, is usually created by the Commercial Analyst inserting risk-related costs or revenues into the first year of the time series of the project. There is no way for the Commercial Analyst to know, at the time of creation of the translation table and before the risks are identified, how to distribute risk impacts over time.

As described previously, the time value of money has a considerable impact on NPV. That is, just for example, the $2,000,000 being considered here, if spent in the first year, would have significantly more (negative) impact on NPV than would the same $2,000,000 being spread out over the last 10 years of the project. Therefore, when the RPP and discipline expert have their conversation, it is essential that the RPP determine which time periods will be effected by the risk impact and what proportion of the impact will be allotted to each effected time period.

Use of Probabilities

I believe it is the common use of Bayesian logic in decision trees that compels people to view probabilities as part of a product involving probability and impact. Typical solution of a decision tree involves using probabilities as "multipliers" for values emanating from leaf nodes. Combination of all leaf-node values and their associated probabilities results in the "expected value" (i.e., actually, not expected) at the left end of the decision tree.

Many times (but not always—as sometimes in the translation-table process), a better practice is to utilize the probabilities as traffic cops. This method is fully explained in "Two Approaches to Solving Decision Trees—A Class-Action-Suit Example" in my book *Risk Modeling for Determining Value and Decision Making*, which is referenced fully at the end of this chapter. Also see the section "Decision Trees Applied to Legal Matters" in Chapter 4 (Figures 4.1 through 4.3 of this book).

Traffic-cop use of probabilities simply means that each time a probability is encountered (in a tree or in the logic of a computer program), a random number—usually between 0 and 1 or 0 and 100—will be generated. The randomly created value is compared against the deterministic probability value or against the probability value selected from a distribution of such values. If the probability represents the chance that an event like a risk *will* happen, then if the randomly selected value is less than or equal to the probability value, the event (risk) is considered to have taken place and all attendant consequences (costs, time delays, revenue increases, etc.) are considered and applied. If the randomly selected value is greater than the probability value, then in the case in which the probability represents the chance that the event *will* happen, the event (risk) is not considered.

Depending on the sophistication of analysis offered to the small groups at the RIW, use of probabilities as "traffic cops" might be possible. For example, in the room where the RIW group meets, there might be a computer available with the economic model loaded on it. It could be arranged with the keeper of the model that two input cells could be made available—one cell for expressing the probability of a risk and another cell for entering an impact value for the risk (sometimes multiple cells are needed to, for example, expressing a dollar figure versus a time delay or advance). The program can then generate in a single output cell a "translated" NPV (common metric) value using the probability as a "traffic cop." The time value of money is not considered at this point because that is a bridge too far for the RIW. If such equipment is not available, then attendees might have to determine the NPV from the simple translation table supplied and use the risk's probability as a multiplier. This method is not preferred, but is sometimes necessary. Such is life.

In the actual risk-monetization process, probabilities are rarely utilized to "damp" the coefficient used to represent an impact. For example, if there is a 50% probability that a $100 fine might be realized, the probability is not generally used to produce a value of $50 (0.5 probability multiplied by $100 impact). This is because in real life, we are going to get fined or not. We will realize $0 or $100. Therefore, in a Monte Carlo model, for example, the 50% probability (0.5) would, on each iteration, be tested against a randomly generated value between 0 and 1. If the randomly generated value fell between 0 and 0.5, then the $100 fine would be realized. If the magnitude of the randomly created coefficient is between 0.5 and 1, then the $100 fine would not be considered in that particular iteration. This is how probabilities will be utilized in the risk-monetization process described here.

Have a Conversation with the Keeper or Builder of the Project Economic Model

In the subsection of Chapter 1 entitled "Holistic Set of Integrated Risks," I address the problem of "hidden" risks. For example, a typical spreadsheet economic model for a project might contain the usual "cast of characters" as line items. These things include parameters such as cost, utilization, production estimates, revenues, and the like. There might not exist, for example, a specific line item dedicated to political problems such as that exemplified by CON 15 or by environmental concerns such as that expressed by ENV 3.

Remember, at the RIW, risks were identified and captured under a small number of high-level categories. Risk CON 15 was captured in the

Construction high-level group and therefore labeled as a construction (CON) risk. Actually, this risk relates to the issue of the local legislature changing the zoning laws in our favor so that we can construct a multilevel parking structure that, under current regulations, is prohibited. Really, this is a political problem.

Likewise, risk ENV 3 represents an environmental problem that, if I know my economic spreadsheet models, will not find a place (specific line) in the spreadsheet. It is essential that such risks be explicitly represented in the project economic model by unique line items. Hiding the impacts associated with these costs in other line items not only bastardizes the costs that are supposed to be represented in the line item (you are "secretly" adding some more cost to an already existing cost, which disguises the already existing cost) but completely hides the impact associated with the risk in question.

So, although it can be a really (really!) awkward conversation, the RPP has to do his or her best to convince the keeper of the project economic model to insert specific lines in the model that address these most important risks (if that is required—sometimes the risk impacts *do* contribute rightly to an already existing line item, but that is not generally the case).

If the RPP is unsuccessful in convincing the keeper or builder of the model that such modifications are necessary, then all I can recommend is that the RPP keep careful track of in which already existing line items the new risks will be hidden. This is an awkward situation especially when the RPP is attempting to explain to management the impact of a risk on the project's value or COS and that risk's impact is integrated with other impacts from which it cannot be separated.

Connect the Economic Model to a Monte Carlo Engine

Although it sounds a simple task to attach to the project economic model (usually a spreadsheet) a software application that will, in conjunction with the model spreadsheet, be capable of at least emulating a Monte Carlo process, such an aspiration is not so easily accomplished. There exist several not-so-easy-to-meet requirements.

First is the issue of integrating COF. In the section "The Inserting Zeros Work-Around" of my book *Risk Assessment and Decision Making in Business and Industry: A Practical Guide*, 2nd Edition, I address this subject in detail. A few words are warranted here.

Each iteration of a Monte Carlo process is an attempt to model the situation independent of other iterations. So, on each iteration, the model is "building a whole new world" so to speak.

Let's consider a project, that is, drilling a water well. There is some chance that the project will never get off the ground. For example, one of the chances of failure might be that, long before we would have started the project in earnest (the actual drilling of the well), we fail to get permission from the land owners to drill on their property. This, in effect, kills the project—abject failure. Of course, there are those readers who will argue that the cost of attempting to get permission to drill is part of the project and, therefore, should be included in any analysis even if the project fails. This is the point at which a business has to decide what is considered simply the cost of doing business—not associated with any particular project—and what costs will be actually attributed to the project itself. It is a line that is drawn in different places in individual businesses. Abject failure typically is defined as failure before significant expenditure on the project has been realized. The project just never gets off the ground.

So, we might estimate that the COF for failure to obtain permission from landowners to drill is 40%. Each iteration of the Monte Carlo process will, then, model a situation in which we do get permission. Remember, if we do not experience abject failure, it simply means that we get to go ahead with the project. The monetary outcome of the water-well-drilling process might not in a given iteration yield a positive result. That is, sometimes we will drill but the flow rates will be too low and our profit margin is unacceptable.

A cumulative frequency plot of profit from the executed project is shown in Figure 8.3. Note that the Y-axis intercept of the cumulative frequency curve represents the 40% probability of abject failure. The probabilities of obtaining any profit level (or more) is impacted by considering this COF. The point to be made here, however, is that the cumulative frequency curve of profit is comprised of 1,000 values. The algorithm did not perform 1,000 iterations and on each iteration test to see if abject failure occurred and, if it did, inserts a zero to represent failure. This process would, first, lead to a stack of around 400 zeros on the X-axis. Worse, the cumulative frequency curve would be comprised of only around 600 values. In situations of a high COF (low COS), the cumulative frequency curve can be comprised of very few values.

It is a reality of random sampling that the range of the cumulative frequency curve—to a point—will be a function of the number of values of which it is comprised. Therefore, the larger the COF, the smaller the range of the cumulative frequency curve. This is not what you want. Addressing COF in the manner shown in Figure 8.3 guarantees that the cumulative frequency

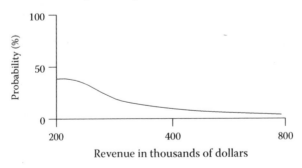

FIGURE 8.3
Cumulative frequency plot of profit with a 40% chance of failure.

curve will always be made up of the full number of values that match the number of iterations.

A second reason that the zero work-around to represent failure is a bad idea is that, many times, zero does not represent failure. For example, if the plot in Figure 8.3 represents the first-year profit of the project, we might fully expect that first-year profits might be negative due to the incorporation of construction costs and other up-front costs in the first-year profit projection. So, the range of profit for the first year might well include zero and such a value does not represent failure. Any Monte Carlo engine to which the project economic model is attached has to be of sufficient sophistication to properly handle COF.

A second reason to carefully select a Monte Carlo engine for attachment to the project economic model is that when addressing finances, it will be necessary to resample a single distribution multiple times—once for each time period, for example—within an iteration. This is a trick that is not easily performed. More on this will be discussed later in this chapter.

Generate the Deterministic Project Base Case NPV

"It's all relative" is an old adage that is aptly applied to risk monetization. It is difficult to make the case that the process of risk monetization results in the betterment of project economics or in an improvement in the chance of project success unless such results can be compared to what might have happened had risk-monetization practices not been applied. Therefore, it is essential that we establish an economic benchmark for the project.

As I have mentioned numerous times in this book, most projects have an associated economic model. Again, in the examples offered in this text, I am utilizing NPV as the critical economic metric. One of the simplest tasks associated with the risk-monetization process is the generation of the "base case" NPV for the project.

Typically, this is accomplished by the project RPP approaching the keeper or feeder of the project economic model (usually the project Commercial Analyst or some other designated individual or group) and requesting that the model—usually a spreadsheet—be run using the most current data. If the model is deterministic, then the resulting NPV will be a single value. If the model is routinely connected to or internally utilizes a Monte Carlo engine, then the mean NPV should be gotten. This NPV value should be plotted on a time-series-like plot such as the one shown in Figure 8.4.

Account for the Risks

Being in the risk business tends to cause one to make every reasonable effort to view and appreciate "the bright side" of life and business. If one did not do this, the preponderance of negative impacts—and the seemingly human trait to focus and dwell on them—might cause no projects to be done at all

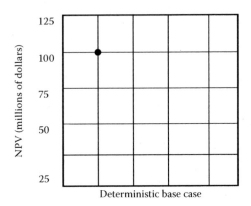

FIGURE 8.4
A time-series-like plot with the project's mean base-case NPV plotted.

(I exaggerate bit, but you know what I mean). So, I thought I'd start this risk-monetization example with a peek at the upside—an opportunity.

In Figure 8.1, risk COM 7 is a commercial risk that relates to the possibility of a sports arena being built sometime in the near future in reasonable proximity to the proposed hotel. Although failure to build the sports arena is not deemed to mean failure of the hotel over the long haul, construction of the arena certainly would boost the proposed utilization rate ("occupancy rate") for the hotel. If the reader has ever seen a real-life spreadsheet economic model for a large project, then it will come as no surprise that such a model (spreadsheet) can't practically be stuffed into a figure in a book such as this. Bowing to this impracticality, I show in Figure 8.5 just one row from such a model.

Utilization is represented as a percent in each of the 15 time periods (years, in this case) of the model. All utilization percentages are those that represent the situation in which the sports arena is *not* built. Management has deemed that this project must "have legs" without the influence of fortuitous events.

First-year occupancy is nil because that is the year of construction. Building is expected to stretch into the second year, which accounts for the low utilization percentage in that period. Utilization percentages after the second year represent projections based on data from existing "similar" hotels.

Conversations between the RPP and the expert for this risk have led the pair to conclude that no fewer than three representations of uncertainty need to be generated to account for this risk in the economic model. A first representation of uncertainty will deal with the probability that the sports arena will be built at all. A distribution will be utilized to relate the effective start year for the arena—that is, when the hotel would realize a positive utilization impact. Another distribution will portray the arena's expected impact on hotel occupancy.

Parameter	Year 1	Year 2	Year 3	Year 4	Year 5	Year 6	Year 7	Year 8	Year 9	Year 10	Year 11	Year 12	Year 13	Year 14	Year 15
Utilization (occupancy) in %	0	25	45	55	60	60	60	60	60	60	60	60	60	60	60

FIGURE 8.5
Taken from the spreadsheet economic model, a single 15-year row of utilization (occupancy) percentages.

Expert opinion indicates that there is a 60% chance that the arena will be built near enough to the hotel to impact its utilization rate. Simple arithmetic would indicate that there is, then, a 40% chance that the arena will not be built, or, that it will be built in a location that does not benefit the hotel. The expert in local affairs can be no more "sure" than representing the probability as a 60/40 split. Therefore, in the economic model, the probability of positive arena impact will be represented as a deterministic value (0.6 in this case). On each iteration of the Monte Carlo process, a random number between 0 and 1 will be generated and compared to the 0.6 value. If the randomly generated number is between 0 and 0.6, then it will be assumed that the arena is built proximal to the hotel. If the randomly generated value is greater than 0.6, the model will assume no impact from the arena on that iteration. In this case, the two representations of uncertainty to be discussed below are moot.

If the sports arena is built (see paragraph above), it is expected to happen no sooner than 2 years from the current date. Because of pressure from the indigenous population to get a new arena built somewhere soon, the starting date for benefiting from arena business is projected to be no later than 5 years from the current date. A distribution of effective years—from 2 to 5—is built (not shown here) to be sampled on each iteration of the Monte Carlo process. For an iteration, a value of 2, 3, 4, or 5 will be selected, and it will be from that time-period forward that the positive impact of the arena will be incorporated.

Deterministic entries (i.e., single values) populate the cells of the spreadsheet economic model. For example, it can be seen that in Figure 8.5, utilization percentages for each time period are represented by single values. Mean (for the year) occupancy rates are low in the second time period due to part of that year spent in construction. Yearly average rates then increase modestly and plateau. Again, these rates represent the "no arena" scenario.

Because events in the proposed arena will be mainly seasonal and episodic, the arena's impact on occupancy rate is expected to be a relatively modest 1%–3% per year. The number of events held at the arena is expected to increase in the arena's first 5 years of operation and then, likely, hold steady. Attendance at arena events also is projected to increase in the first 5 years of operation and then plateau. Therefore, the 1%–3% utilization rate for the hotel will be cumulative for a 5-year period and then, like arena events and attendance, will plateau.

A scenario such as the one described above might be perceived as relatively simple to represent in a probabilistic model, but most commercial Monte Carlo engines that attach to spreadsheets are poorly prepared to handle this situation. Typically, a Monte Carlo process will sample each input distribution one time in iteration and use the selected values to calculate the results for that iteration. Introduction of multiple time periods significantly alters the demands upon the Monte Carlo system.

For example, in our model, we are assuming a 1%–3% cumulative increase in hotel utilization during the first 5 years of arena operation. If we follow the standard Monte Carlo methodology and sample each distribution just once during a particular iteration, then we might, on an iteration, select a utilization increase of, say, 3%. The model is comprised of multiple time periods to which the percent increase needs to be applied. If the software restricts sampling of a distribution to one time per iteration, then only the 3% value will be available for each time period. When we indicated that there would be a 1%–3% cumulative annual increase in utilization, did we really mean that, if 3% was selected, the increase would be 3% for each of the 5 years? I think not.

What we are trying to indicate is that in any given year, there could be a 1%–3% increase in utilization relative to the previous year (cumulative increase). Given that we are allowed only one sampling of the utilization-increase distribution on any given Monte Carlo iteration, then we are stuck with the selected value for each time period. This, I think, is not what is intended. The "solution" of creating a separate utilization-increase distribution for each time period is not a practical approach. Impracticality stems mainly from the preponderance of parameters that are represented by time series.

Without the ability to resample a distribution—once for each time period—in any iteration of the Monte Carlo model, the envelope that represents the range of utilization increase will represent the impact of experiencing a near 3% increase each year. Such a line is indicated in Figure 8.6.

Given that we expressed a 1%–3% annual cumulative increase in utilization, we would expect that, just for example, we might experience a 2.8% increase in year 1 of the arena impact but a 1.1% increase in year 2, a 2.0 increase in year 3, and so on. The cumulative impact of such increases would be represented by a yearly-increase line such as that shown in Figure 8.6. In order to construct such a line, however, resampling of the utilization-increase distribution for each time period (within a single iteration) is required. Further complicating matters might be the requirement to restrict year-to-year swings in utilization rate, but that is an entirely different conversation and I won't address that here. The plot in Figure 8.6 is stylized in that all increases are shown starting in year 2. In the real model, the increase could start in any of 5 years.

So, assuming we have available the sophistication to resample the utilization rate for each time period within each Monte Carlo iteration, the model would on each iteration:

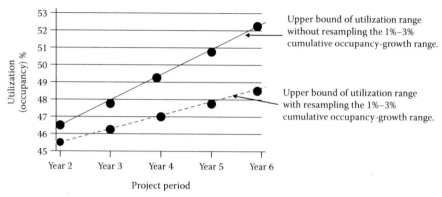

The above plot shows growth beginning in
year 2. As explained in the text, growth might
begin in year 2, 3, 4, or 5 and continue for
5 years before reaching a plateau.

FIGURE 8.6
Envelopes of utilization contrasting the capability to resample a distribution once per time
period within a Monte Carlo iteration (solid line) with the inability (dashed line) to resample the utilization distribution for each time period within a single iteration. All utilization
increases here are shown for the 2nd year only.

- Test to determine whether or not the arena is built.
- Determine the year in which the impact of the arena is realized by
  the hotel.
- Resample the utilization-increase distribution for each time period
  to determine the utilization rate for the time period.

Adding just this risk (opportunity) to the probabilistic execution of the economic model, the resulting (one per iteration) NPV values are calculated and
stored as a distribution. This distribution of values will be utilized and displayed in an expansion of the plot shown in Figure 8.4. Such a plot will be
shown later in this chapter.

Next, I will address risk COM 27. In the section of this chapter entitled
"Description of the Risk Monetization Example," risk COM 27 is defined
thus:

> If another competitive hotel chain is successful in obtaining large tracts
> of already-in-use land in the area, then labor rates will become more
> competitive and will impact revenues and NPV. If, because of the build-
> ing of the competitor hotel, our labor rates are projected to be prohibi-
> tively high, then our project—if in its initial stages—might be cancelled.

It is at this point that I have to point out to the user that the hotel example
being utilized here is completely contrived. That is, I could have designed it
any way I wished. I also have to admit that regardless of how complex this

entire risk-monetization process might seem to the reader, I have been presenting a sort of "la la land" version of reality.

Risk COM 27 is going to be a bit of a wake-up call. I have purposely designed it and its integration into the monetization process to be more complex. That is, it will take some "head scratching" to determine just how to integrate this risk into the overall monetization scheme. I will present one way to approach integration of this threat. The reader is, of course, free to disagree with my approach and to formulate his or her own opinion and method for bringing risk COM 27 into the fold.

First, I would like to point out that most of the individuals at the RIW have little or no idea how the project economic model actually works. So, it is not uncommon for attendees to generate risks the impacts for which will not neatly fit into the general design of the economic model. This is separate and distinct from the "no line item" issue previously discussed.

Threat COM 27 addresses two aspects of the risk assessment—labor costs and COF. In the section of this chapter entitled "Types of Chance of Failure," the distribution to be used to represent labor costs is shown in Figure 8.2. The reader is referred to the aforementioned section of this chapter for a full explanation of the distribution and how it relates to COF.

Labor rates in Figure 8.2 are expressed as a range. Not only the range results from management's uncertainty regarding exactly what the per-hour cost per employee might be in a noncompetitive situation (a competing hotel is not built), but is influenced greatly by the quality of the competitive hotel.

If a cut-rate hotel is built, then the labor rate is likely to stay below the cut-off value shown in Figure 8.2. These are rates management can "live with." However, if a luxury-brand hotel is built, then because of competition for the projected sparse labor force in the area, rates might exceed the projected acceptable limit (cutoff value in Figure 8.2). If in its preconstruction stages when a luxury-brand hotel begins to acquire property, the project might be cancelled. It is believed by hotel management that if they break ground first, construction of any type of competing hotel in the area will be exceedingly unlikely.

As can be seen in the partial risk register in Figure 8.1, representatives at the RIW believed that there is only a 10% chance that a competing hotel will be built within the 15-year timeframe of the economic model. Figure 8.2 indicates that about 18% of the values that comprise the labor-cost distribution lie above the cutoff value.

Given that the project might realize hourly labor rates greater than the cutoff value only in 10% of the Monte Carlo iterations, the COF value (18%) will be impacted by the 10% probability before being combined with other COF values (in this simple example, only the COF from risk ENV 3) to determine the Y-axis intercept for the NPV cumulative frequency curve. But that's not all.

It is further believed that if (10% probability) a competing hotel is built, there is a 50/50 chance that competitors will begin their land acquisition process before our hotel-construction phase has started. If that is so, then our hotel would be cancelled and that effect would be indicated by the COF associated with risk COM 27. So, the COF to be used in the COF-integration process would be further "damped" by the aforementioned 50% probability of a competing hotel being introduced early in or hotel's project lifecycle.

If the competing hotel is started after we have commenced construction on our hotel, then our project might incur hourly labor costs that are in excess of what management deems acceptable. Again, there is a 10% chance of a competing hotel being built at all and a 50% chance that if it is built, it will occur after it is practically too late to call off construction of our hotel.

Programming within the economic model is designed to draw a value from the "good" (less than the cutoff value) section of the distribution shown in Figure 8.2 and utilize that value as the initial-year labor-cost value. The programming automatically increases the labor costs by a factor that is tied to an estimate of annual inflation, so the value drawn for labor costs from the distribution shown in Figure 8.2 does not have to be modified for each time period as was true for the utilization rates discussed in the preceding risk discussion. This does not mean, however, that the impact of labor costs is not considered on a time-period basis. Unlike the simple translation table described in previous chapters to roughly translate the impact of a risk to a common metric, the time value of money certainly will be considered here.

OK, here's the complicating issue I am intentionally inserting into the process. Folks at the RIW, as previously stated, don't have an intimate understanding of how the economic model operates. They do know, however, what is the acceptable range for hourly labor costs. Given that they believe that they are attending the RIW to identify and quantify threats and opportunities, they decide to represent the impact of this threat as the consequence that will be realized if acceptable labor rates are exceeded.

Values, therefore, shown as "Impact Minimum," "Impact Peak," and "Impact Maximum" in Figure 8.1 for risk COM 27 represent the difference between our upper limit of per-hour labor costs ($13) and the costs that exceed our upper limit ($14, $15, and $16). This difference ($1–$3) is an hourly figure that, by the members of the RIW, was multiplied by the applicable number of employees and hours per year. This product represents the values that are a threat to the project. In the economic model, these values would be appropriately impacted by the 10% chance that the competing hotel would be built and by the 50% chance that if built, it would materialize too late for our hotel to be cancelled—thereby leaving our project exposed to the potential of high labor costs. A point to be made here is that as the RPP, you have little control over how attendees at the RIW decide to represent the risks.

Now, I know that some of you are thinking that we might be "double dipping" failure by using the values above the cutoff to represent abject failure and then to use them again to represent the economic impact on the project. If you are thinking that, then it demonstrates that you are gaining insight into the process.

I would argue that we are not double-dipping failure in this example. The COF (18%) calculated as a percent of the values that lie above the cutoff is then impacted by the 10% likelihood that the risk will materialize and also by the 50% chance that even if it does materialize, the timing of the construction of the competing hotel will be, relative to our hotel's schedule, soon enough to cause our project to be cancelled. So, the COF from this risk will (only slightly) impact the Y-axis intercept of the NPV cumulative frequency curve. The point at which the curve intercepts the Y-axis indicates the probability that the project will go ahead (COS) or be cancelled (COF). The COF to which the COF for COM 27 contributes represents the situation in which the project is cancelled.

If the project is not cancelled—the "alternate" world, then there is some chance that a competing hotel will, in fact, be built, but too late for us to cancel our project. The X-axis values of the NPV cumulative frequency plot will represent the situation in which we do build the hotel. It is appropriate to consider the NPV range for the project impacted by the probability-affected hourly-labor costs that exceed our management's acceptable limit.

OK, so what's the problem? The problem is that the already existing economic model likely has a cell or time-series of cells that address hourly labor costs. The model is currently designed to draw from the "acceptable range" of labor costs. Attendees of the RIW have decided to represent this threat, as described above, as a difference between the acceptable upper limit and the range of "unacceptable" values—those differences being multiplied by the number of hours per year and a guess at the number of applicable employees. How can the RPP get this threat—expressed as it is—into the existing economic model? One approach, and it is the one that will be taken here, is to convince the keeper of the economic model that a new parameter needs to be added to the model to represent these costs and to impact these costs appropriately with the attendant probabilities and to apportion the costs to the appropriate time periods. Of course, there are other ways of approaching this, and the reader is free to conjure up alternatives.

I purposely designed risk COM 27 to be "messy." Both the COF and the hourly costs that exceed management's upper limit (cutoff) have to be dealt with. As you can see from the preceding discourse, it is not always easy to take account of threats and opportunities—especially when individuals and groups at the RIW are free to express risks as they see fit. Again, I present above just one approach to integrating this risk.

After applying the utilization-rate opportunity to the economic model and recording its impact, that opportunity can be removed from the model and the labor-cost distribution (threat) should be implemented. The relative consequence of labor costs will be shown in a figure below that is an expansion of Figure 8.4. The tiny COF associated with this threat will be integrated with the COF from risk ENV 3 (to be discussed below) to impact the Y-axis intercept of the NPV cumulative frequency curve. An example of such an impact is shown in Figure 8.3.

Next, we apply risk CON 15 to the economic model. This risk is described as follows:

> If zoning laws are not changed so that a multi-story parking facility can be built, then additional already-in-use land will have to be purchased for parking. This will significantly negatively impact the first-year CAPEX and, as a result, negatively impact NPV.

This risk has associated with it no COF and is simply represented by two uncertainty ranges. One is a range of probability. The other is a range of costs that will be added to the CAPEX in and only in the first year of the project.

As with the utilization opportunity, a probability range will be established by conversation between the RPP and the discipline expert for this risk. Let's assume that they settle on a range of 20%–60% likelihood that the zoning laws will be changed (distribution not shown here). Then, on each iteration of the Monte Carlo process, a value between 20 and 60 will be selected—let's say on iteration 1 the value selected is 30 (or, 0.3). A random number between 0 and 1 is generated by the computer model. If the randomly selected value is less than or equal to 0.3, then no costs are added to the first-year CAPEX (because the zoning laws were changed). If, however, the randomly generated value is greater than 0.3, then first-year CAPEX will be impacted by costs associated with land purchase.

It follows, then, that the second distribution associated with this risk is one that represents the first-year CAPEX associated with purchasing land that will substitute for the multilevel parking garage that is disallowed by zoning laws. If warranted by the probability test described in the preceding paragraph, a value will be randomly selected from the land-purchase-cost distribution (not shown here) and that cost added to the first-year CAPEX in the economic model. Just like the other risks described thus far, the process delineated above is repeated on each Monte Carlo iteration and the resulting NPV recorded. The resulting NPVs constitute a range and that range will be represented in the figure below that is an expansion of Figure 8.4.

This leaves risk ENV 3. This risk is stated as follows:

> Risk ID ENV 3: Water runoff from the paved hotel property will exceed current acceptable limits. If a variance permit is not granted, then the entire project will be cancelled.

A high ranking of this risk was deemed warranted because it represents a high-likelihood threat to the entire project. This threat is a COF-only risk. That is, it will impact only the *Y*-axis intercept of the cumulative frequency curve and will not be represented as a term in any of the equations that calculate NPV. It will, however, impact the likelihood of the project attaining any given NPV value. This influence will be demonstrated in a subsequent figure in this chapter.

Viewing and Interpreting the Results of the Risk-Monetization Process

I have in previous sections of this chapter referred to plots that will be an expansion of the plot in Figure 8.4. Such plots are shown in Figure 8.7.

Shown in Figure 8.7 are variations on the "high/low/close" chart. Plots on the far left of Figure 8.7—one for each of the NPV-impacting risks—show the deterministic (or mean, if the original spreadsheet model was driven by

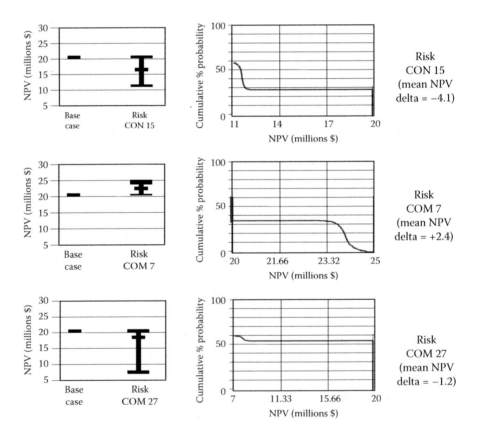

FIGURE 8.7
Leftmost plots are a variation on the stock broker's "high/low/close-type" plot showing the base-case NPV (from Figure 8.4) and the range and mean (middle tick mark) probability-impacted NPV of three of the four risks utilized in the hotel-project example. Rightmost plots are the actual cumulative frequency curves that are represented by the vertical bars in the leftmost plots. Delta NPV values are the difference between the deterministic base-case value ($20 Million) and the mean NPV of the risk's cumulative frequency range.

a Monte Carlo process) NPV for the project base-case scenario. This is the situation without impact from any risks identified at the RIW.

Plotted immediately to the right of the base-case NPV is a "bar" representing the range of NPV that resulted from applying the risk in the economic model. The horizontal cross bar that is not at the top or bottom of the bar represents the mean of the NPV range. Note that because risk COM 7 represents an opportunity, the mean NPV rises relative to the base-case NPV. The "value" of this risk is shown by the difference between the base-case NPV and the mean NPV resulting from application of this risk. The other two threats—risks CON 15 and COM 27—both diminish the NPV. The "values" for these risks are determined in an identical manner to that of risk COM 7.

In Figure 8.7, uncertainty is represented by the "length" of the vertical bar in each plot. For each risk, the cumulative frequency curve that the vertical bar represents is shown in the three plots to the right of the "high/low/close" graphs. On these plots, the impact of COF can be seen. Probabilities of attaining any NPV value (or greater) is significantly affected by the COF.

In Table 8.1 is shown the impact of each of the risks. The "NPV impact" is simply the change in mean NPV (see Figure 8.7) caused by incorporation of the risk relative to the base-case value. Given that the project team has chosen NPV as the "answer" metric, it is up to the project management to determine just how much CAPEX or operating expenditure (OPEX) is justified—and over which time periods—to mitigate the threat or capture the opportunity. Clearly, if the cost of mitigation or capture is deemed excessive relative to the NPV impact for a risk, then mitigation or capture actions might not be enacted. The absolute cost of mitigation or capture action and the timing of that action will help determine whether or not the action is economically justified. Any other "answer" metric can be plotted.

Note that the NPV values shown in Table 8.1 are somewhat different than those shown in the "Translated Impact (NPV n $MM)" column in Figure 8.1. The difference is caused by the rough estimate of impact generated by the

TABLE 8.1

The Mean Probability-Impacted NPV Impacts—
Relative to the Base-Case Value of $20 Million—
for the Two Threats and One Opportunity
Considered in the Hotel-Project Model

Risk ID	NPV Impact (Millions $)
COM 7	+2.4
COM 27	−1.2
CON 15	−4.1

Note: Impact values for each risk were properly discounted and distributed in the time series for each risk, thus accounting for the time value of money.

translation-table method, by the RPP and discipline expert reviewing the top-ranked risk in a post-RIW comprehensive conversation during which new or different information might be offered, and by the accounting for the discounted time value of money in the economic model (spreading the risk impact over the appropriate time periods and applying a discount rate). I have purposely shown only slight differences between the two calculations of NPV, however, in actual models, it is not uncommon to realize significant changes in value with regard to those values generated at the RIW and those resulting from the full risk-monetization process.

Without information such as that conveyed in Table 8.1, project teams can waste inordinate amounts of time and money "fixing" risks that will have relatively little impact on the project and, perhaps worse, will not address salient risks because they do not realize the real-world consequence associated with a risk. Following a risk-monetization process as described here is absolutely essential if success and profitability are to be maximized.

In previous sections of this book and in other books of mine, I have advocated the integration of risks. In the hotel example in this chapter, I have made example of assessing independently the impact of each risk on the base case NPV. That is, apply one risk, record its impact, and then remove it from the economic model before assessing the impact of a subsequent risk.

It certainly is possible to implement all risks simultaneously to see the integrated impact on the project's value. I do this regularly. However, in a real-world risk-monetization effort, it is not so straightforward as to how to combine all of the top-ranked risks.

In the section "It's about Time" of Chapter 1, I relate a story regarding project-start time. In that narrative, I demonstrate that resources to mitigate threats or capture opportunities must be taken into account. For example, a project team might be planning to enact threat-mitigation actions for N risks in the first 6 months of the project. However, half of those mitigation actions might independently completely consume the time and resources of a critical discipline. It should be obvious, then, that not more than one mitigation action can be enacted at any one time (mitigation actions have to be performed in "series" rather than in "parallel" mode). So much for the 6-month timeframe!

So, what's my point? My point is that it is not just a matter of throwing all of the threats and opportunities into the economic model at one time and seeing what integrated impact looks like. Much homework must be done to determine just which risks and mitigation or capture actions can be simultaneously considered. There is no easy way out of this except to have the RPP and helpers determine which risks might be initially combined in the economic model. If the economic model is sophisticated enough to handle risks and mitigation or capture actions "in series," then that's great. However, that's not typical.

If the economic model is not originally designed to be able to account for risks and mitigation or capture actions in "series," it can be a nightmare of a programming task to attempt to ensure that such integration—inside a complex economic model—is properly executed. It certainly is insightful to realize the integrated impact, and I would be glad to advise any readers regarding how I have approached this in the past. Because of the "one off" (unique) nature of economic models and risk registers, it is far beyond the scope of this book to address the full-monetization integration of all risks. Again, feel free to contact the author for advice on this matter.

But take heart! If you have executed the risk-monetization process as described in this chapter, then you have, in a manner, performed integration of the disciplines. In Figure 1.1 of Chapter 1, I illustrate that it is essential to take account of risks from all major disciplines if you are to have a chance at determining the real-world value and COS of a project. By bringing all of the risks together into one risk register, by transforming their probability-affected impacts into a common metric (I used NPV in the example), and by ranking the risks relative to one another you have, in effect, taken an integrated (holistic) approach to the problem. Believe me, even without actually integrating the impacts of the risks in the economic model (and that should be done if it is practical to do so), the data shown in Table 8.1 will be startling enough to drive the right actions by decision makers.

As described in the section "Sensitivity" of Chapter 5 and depicted in Figure 5.5, a sensitivity plot can be generated for a risk-monetization model. Such a plot will indicate which input parameters—represented by distributions—are contributing most (and least) to the uncertainty of the "answer" variable. This information can focus project teams on the parameters that might increase certainty if effort were expended, for example, to collect more accurate or precise data on the parameter in question.

Although this might seem like the end of the road for risk monetization, it is really just the beginning of the effort for the RPP. Remember, as pointed out in Chapter 6, it is the responsibility of the RPP to regularly review the risk register, regularly capture new risks, pass judgment on the quality and completeness of mitigation or capture actions that people claim to be accomplished, regularly report to the Project Manager and other decision makers, and to perform a whole host of other ongoing duties. Monetizing the initial set of risks is just a mile marker along the way.

Well, that's the story of risk monetization. As should be obvious, the monetization process is primarily about preparation of the risks—and the attendant organization—for the actual calculation of the risk-specific impacts on project value and success. Best of luck with your implementation of the risk-monetization process and feel free to contact me if you need help.

Suggested Readings

In addition to books and articles cited in the preceding text, this list includes other material that the reader might find helpful and relevant to the subject matter discussed in this section of the book.

Brockwell, P. J., and R. A. Davis. *Introduction to Time Series Analysis*. 2nd ed. New York: Springer-Verlag, 2002.

Chatfield, C. *The Analysis of Time Series*. 6th ed. Boca Raton, FL: Chapman & Hall/CRC Press, 2003.

Gentle, J. E. *Random Number Generation and Monte Carlo Methods*. 2nd ed. New York: Springer-Verlag, 2003.

Jaeckel, P. *Monte Carlo Methods in Finance*. New York: John Wiley & Sons, 2002.

Koller, G. R. *Risk Modeling for Determining Value and Decision Making*. Boca Raton, FL: Chapman & Hall/CRC Press, 2000.

Koller, G. R. *Risk Assessment and Decision Making in Business and Industry: A Practical Guide*. 2nd ed. Boca Raton, FL: Chapman & Hall/CRC Press, 2005.

Vose, D. *Risk Analysis: A Quantitative Guide*. 2nd ed. Chichester, UK: John Wiley & Sons, 2000.

Appendix

*Summary of the Journey from Risk
Assessment to Monetization and More*

Section 1: It's All about Changing Behaviors

When I first envisioned penning the summary that comprises this appendix, I was quite enthused. Inspiration stemmed mainly from the prospect of being able to offer a summary of the major points in the journey from risk assessment to monetization, to management to implementation. However, my enthusiasm was damped by the sheer preponderance of messages to be condensed and delivered.

It was clear that anything resembling a credible job could only be done through a series of focused sections. Section one relates to behavioral change. Some of you might be looking for the conveyance of risk-related technical practices in these notes. You won't be disappointed—such techniques will be addressed in later sections.

Where to begin? Given that I will be discussing the many facets of risk, it seemed prudent to first define the term—"risk," that is.

In business, if you ask N people for their definition of risk, you likely will get N unique responses. Within any corporation, each discipline views risk through a colloquial lens. For example, the Health and Safety department will define risks typically as threats to be rooted out and eradicated. Denizens of the Finance department, however, view risk as a positive thing. They seek and embrace risk (hopefully, not more than they can handle) because their job is to maximize return—low risk, low return; higher risk, higher reward. To them risk is an opportunity.

So, the definition that I propose for "risk" is

"A pertinent event for which there is a textual description."

The "pertinent" term indicates that a risk is something that has a material impact—either positive or negative—on, for example, project value. Therefore, I refer to risks as either threats or opportunities. A risk is an event. That is, it is something that might happen. This event can be textually described (you can tell someone what it is).

Associated with each risk are at least two other parameters: probability and consequence (or impact). About both probability and consequence, we can be either sure or uncertain. To learn more about this, I refer you to the books listed at the end of this appendix.

After many years in the academic, government, and corporate arenas— and even in spite of my 20-some-odd years of research experience—I have come to realize that implementation of practical, relevant, and effective risk assessment/risk management (RA/RM) practices is not mainly about new technologies and so on—it is primarily about changing behaviors. Changing behaviors, in turn, is mainly dependent on the reward system.

With regard to implementing RA/RM practices in project teams, just what are we (those of us who promote these things) proposing project teams do? A basic set of recommended processes would include the following:

- Hold a facilitated early in the project risk-identification event.
- Record risks and other information in a risk register.
- Regularly review existing risks and identify new risks.
- Take a holistic approach—that is, risks identified should include those from health and safety, security, medical, legal, logistic, engineering, scientific, country, financial, commercial, and other areas.
- Record in the risk register mitigation-of-threat/capture-of-opportunities actions. The term "mitigation," in this context, no longer relates to "firefighting" (what to do if the risk occurs) but, rather, refers to actions to be taken *early* with the aim of preventing the threat from ever materializing, or, to attempt to ensure capture of the opportunity.
- Assign a risk-process proponent (RPP) (a person) who will shepherd the RA/RM processes throughout the life of the project.

So, if you were a project team leader, why would you want to do this? Given the typical reward system, you are rewarded for successfully launching a project, but not necessarily for launching a successful project. In addition, you are in competition with other project leaders for monetary and human resources. Why would you want to identify most of the risks for your project (shine a bright light on your project's warts) when the other guy is not doing this?

In addition, what could compel you to spend money now to mitigate a threat, for example, (and remember, "mitigate" means taking steps to prevent the threat from materializing) that might or might not happen? Even though, later "fixing" the impact of the risk will undoubtedly cost more money than the preventative action will cost now, why should you not just take the chance that the risk will not happen? After all, you get rewarded for saving money and you likely will have moved on by the time the risk might materialize.

So, as the RPP, what argument/proof would you offer to cause a project team to WANT to implement this process? After all, in the past you have had project team leaders wag their fingers in your face and say: "Had you not chased me around with all this risk stuff, I would have been in a better place faster and cheaper—all you did was waste my time, money, and effort!" And, by the way, "Well, oh yeah?" is not an appropriate response. On projects that take years to go from inception to fruition, what case can you make that might convince any project team that implementation of a cogent RA/RM process is well worth their while? In spite of the existing reward system, what can you do or say that will change their behaviors? In Section 2, we will explore some first steps to take to begin to change behaviors.

Section 2: Some First Steps to Begin to Change Behaviors

The last section ended with the project team leader wagging his/her finger in my (the RA/RM proponent) face and accusing me of wasting his/her time, money, and effort with this risk stuff. The section concluded with the queries: "On projects that take years to go from inception to fruition, what case can you make that might convince any project team that implementation of a cogent RA/RM process is well worth their while? In spite of the existing reward system, what can you do or say that will change their behaviors?"

If I Hadn't Believed It, I Wouldn't Have Seen It

Far be it for a guy like me—and you'd have to know me to realize how true this is—to attempt to convert you to "that old time religion," but I might as well admit right upfront that at least some of the perceived "good" that emanates from a practical and well-implemented RA/RM process is a matter of faith. On long-term projects (years to complete), it really is partly a matter of "If I hadn't believed it, I wouldn't have seen it." I hate to admit it, but it's true.

Years ago, when the world was young, I worked in research support of the oil-and-gas exploration arm of a major energy company. When I say "exploration" here, I mean actual "wild cat" well drilling (drilling in completely unexplored/untested areas) and not already-discovered field infill drilling, and so on. Every year, the exploration portfolio would contain on the order of 30–50 proposed wells worldwide. In a given year, the company might actually drill and realize the results from, say, about 20–30 of those wells. So, from the time someone got sanction to put a proposed project into the exploration portfolio to the time we actually found out the results of the drilling (whether or not it was an economically successful venture), the elapsed time was, about, 1–2 years. In the energy business, this is really rapid-fire project execution. By the way, at that time, a "good" chance of *economic* success for any given exploration well ranged from around 10% to around 35%—that is, there was a pretty good chance that the exploration well would not be economically successful. This was generally true throughout the industry at that time.

Because of the yearly completion of, say, two dozen exploration projects, over the years it could be demonstrated that the success rate for exploration wells was low. To make a long story sort of short, when we implemented a consistent and holistic risk-assessment process, it could be demonstrated that the success rate of exploration wells improved significantly. The positive impact of the risk process could be without-a-doubt touted because

- Nothing else had changed significantly.
- The real-world results from these short-term projects could be calculated/documented.
- Our prediction of which projects would be most successful improved dramatically.

Improvement could only be demonstrated, however, because within a few years the great number of short-term projects allowed us to compile statistics about improved project success. Because nothing else had changed significantly in that same time period (no other new techniques had been introduced that could "steal the thunder" from the risk process), it was impractical to argue that implementation of the risk-assessment process had not dramatically improved our fortunes.

Separate Personal Success from Project Success

Well, isn't that nice? How fortunate were we to have been working on a portfolio of short-term projects, the results from which could be unequivocally known and documented? For portfolios of such short-term projects in any business, demonstration of a positive impact of RA/RM will be similar. But portfolios of such projects are not the norm. Most projects take years—sometimes a decade or more—from start to "finish." Often, at the "end" of such projects, it is difficult to demonstrate/document whether or not the project had been a success. Over long time periods, there surely were many personnel changes, process introductions and removals, changes in scope, and so on. So, in such situations, the perceived benefits from RA/RM process implementation begin to fall into the realm of faith.

Those of us who believe in the RA/RM process can't imagine how it could fail to have a positive impact. However, if you are not a member of the choir, and you—a project manager for example—are charged with launching a project on time and within budget (and, of course, the schedule and budget both get "cut" as we go along), what might compel you to *want* to implement an RA/RM process? Remember, you (the project manager) are getting rewarded for successfully launching the project—this means on time and on budget. So, when the RA/RM proponent comes along and asks you to identify all of your project warts, record them for everyone to see, and implement threat-mitigation/opportunity-capture plans that would cause the threats to be less likely to materialize and opportunities more likely to be captured (i.e., spend money now on things that might or might not happen), why would you want to demonstrate such behavior? This, I think, is where we were at the beginning of this section—ah, nothing like progress!

The reward system is a primary driver of behavior. Sure, morals, ethics and such play a part, but the reward system is undeniably a major influence on behavior. So, the question becomes: How can we get the reward system to work in our favor—to encourage the use of an RA/RM process.

Just one approach is to separate a corporate employee's sense of security from the perceived success of a particular project. I always use the example of the assembly-line worker in a fictitious auto factory. Let's say that the worker on the assembly line linked his/her job security to the sale of one particular car. If the worker saw serious flaws in that single car, he/she might be tempted to overlook or attempt to downplay flaws that could not readily be remedied so that the car would sell. In the project world, this is equivalent to a project team member linking their personal fortunes to the success of a particular project. This philosophy seems strange for the example of the car factory assembly-line worker, but it can be the prevailing attitude of corporate project personnel.

Contrast this with the assembly-line worker who, more realistically, links his/her job security to the quality of the portfolio of cars that roll-out of the factory. In this case, the worker knows that if the public perceives the line of cars to be of low quality, they will shop elsewhere and the car line and all of its workers will fall on hard times. In this case, the assembly-line worker is likely to point out flaws in individual cars so that a portfolio of the best cars they can build goes out the door. In the project world, this translates to a corporate project team member who links his/her long-term employment not with a particular project, but with the best portfolio of projects the corporation can assemble.

In the best-portfolio scenario, the reward system promotes individuals who "call 'em like they see 'em" with regard to threats and opportunities (risks) associated with a particular project. This philosophy aligns perfectly with the RA/RM process, which promotes early identification of probable threats and opportunities and the establishment of threat-mitigation and opportunity-capture plans that will minimize the materialization of threats and the realization of opportunities.

So, how can we create an organization and a culture that promotes the best-portfolio reward system? That is, how do we encourage employees to take an objective view of their project and to want to spend money now to address potential (probable) threats and opportunities rather than attaining "hero" status later on by beating realized threats into submission with big bags of cash (i.e., firefighting)? This, of course, is outlined in detail (like this section, in layman's language) mainly in the latest of the three books referenced at the end of this appendix and will be the focus of the next few sections. Eventually, I'll get to the technical stuff.

Section 3: Communication

We concluded the second section with the following queries:

- How can we create an organization and a culture that promotes the best-portfolio reward system?
- How do we encourage employees to take an objective view of their project?
- How do we discourage the firefighting mindset that requires big bags of cash when threats "blow up" instead of making relatively small investments when the threats "show up"?

I have previously alluded to the fact that changes in the organization and culture require changes in the reward system. The reward system, however, is just the salient issue in a relatively long list of impediments to positive change. Other less potent but equally diabolical considerations include

- Language (see below).
- Inability to implement (we know what to do but can't make it happen).
- Internal competition (enough said).
- Value bias (we prefer to optimize the base case).
- Desired ignorance (just don't want to see the dark side).
- Lack of monetization (can't or won't convert risk into impact on value).
- Ignoring "soft" risks (ignoring political, cultural, etc. type risks).
- Lack of challenge (I won't hurt you if you don't hurt me).
- Lax accountability (we just don't want to accept responsibility).
- Deterministic mindset (just give me THE number!).
- Pressure from "the Street" (it's all about the next quarter).
- Failure to recognize/exploit unconventional relationships.
- Pressure to sanction projects (political impetus to sanction projects).
- It's nobody's job (I'm not responsible).
- Pushing responsibility to lower levels (someone below me will handle it).

This appendix of 10 sections precludes addressing all of these items (although they are addressed in detail in my 2007 book—see references at the end of this appendix). However, I would be remiss in my duties if I did not at least address the second most insidious issue—that of language.

Although technical understanding and capability are critical to success, such prowess pales in importance relative to an in-depth understanding of organizational and cultural aspects. See Koller (2000, 2005) to gain a better understanding of technical aspects. So, while it is my desire to launch into a technical tirade, there are some more seemingly (but not actually) mundane issues to address—not the least of which is language.

To change a culture—that is, to address the three questions at the top of this section—the ability to communicate is paramount. Use of common terms and definitions are at the core of communication. It is folly to attempt to implement an organization-wide metamorphosis without first creating a common language. I can't relate how many times I have noted problems of significant magnitude that were directly the result of miscommunication and misunderstanding.

In the first of this series of sections, I defined the term "risk" as "A pertinent event for which there is a textual description."

This event (risk) can have a positive (opportunity) or negative (threat) impact on the value of a project and is probable in nature. This definition, however, likely will contrast or even conflict with a colloquial definition embraced by denizens of an enclave of the corporation. For example, some folks might adhere to the concept that "risk" is the product of probability and impact. There can be as many definitions of the term "risk" (and of other terms) as groups surveyed.

I have found it folly to attempt to change the way people define risk in their area. For members of the various corporate cul-de-sacs, the definition they prefer makes sense for their business. However, it is still true that because we need to address risk in a holistic way—that is, consider the legal, environmental, technical, commercial, financial, engineering, logistical, security, and other aspects—it is essential that all of the pertinent disciplines be able to effectively communicate. Just as important is consistent "vertical" communication to higher ranks in the organization.

What to do? Because of my German heritage, in high school and college I decided that studying the German language would be a good idea (the jury is still out). In those studies and in a later visit to Germany, I came to realize that there exist many significantly different dialects. To facilitate countrywide communication, newspapers and the like utilize a common "dialect" referred to as "Standard German." I decided that this means of creating a common language would be most effective in risk communication.

So, rather than attempt to change the way each area views and defines risk, I decided that I would promote a common definition of risk (and other terms). That is, when communicating between factions or communicating "up the chain," corporate employees would utilize the common definitions. In that way, regardless of the language used in any area of the corporation, communiqués issued for consumption outside that area would be clear. This works.

While coming to consensus regarding the definition of the term "risk" is critical, so is agreement on a plethora of other terms and concepts. For example, prior to the implementation of the holistic and early-in-project-life risk-identification process, the term "mitigate" typically referred to what should be done when (after) a threat materialized (when the wheel comes off the car, what do we do about that?). When the early risk-identification process is used, however, the definition of "mitigate" changes to mean what can be done now to try to ensure that the threat never materializes (I have identified the wheel coming off as a risk. What can I do now, before the car trip, to try to ensure that the wheel does not come off the car?). Responses to these two views of mitigation are, obviously, necessarily very different. The first is termed "firefighting," which is relatively expensive and disruptive and is to be discouraged. The second is proactive, but is hard to "sell" to organizations because it requires them to spend money now on something that might or might not ever happen.

Alignment should be reached on myriad critical terms and definitions. You will need to address the use of the terms "probability" (the likelihood that an event will happen even once) versus the seemingly interchangeable term "frequency" (the number of times something has happened in the past, or, is forecast to happen in the future), the difference between a "risk" and a "hazard," and a host of other terms and concepts. I can't stress enough the importance of alignment—both "horizontal" and "vertical"—regarding language. Given the holistic approach advocated (considering law, security, commercial, financial, technical, engineering, logistical, and other aspects), I also can't stress enough how elusive alignment can be.

So, in this section, I have addressed one salient aspect related to the query: "How can we create an organization and a culture that promotes the best-portfolio reward system?"

In the next section, we will together explore some practical steps you can take to address the other two questions with which I closed Section 2 and opened this section.

Section 4: Metrics

In each of the three previous sections, I promised to eventually address the more technical aspects of risk assessment and risk management. In this section, I will inch toward a discussion of technical details, but readers should be assured that even the most erudite technical concepts and practices will be presented in plain, understandable, layman's language.

In Section 3, I mainly addressed the arena of risk-related language and communication. Given that any risk-based conversation is likely to include numerical components—discussions of probabilities, costs, schedules, revenues, and the like—it is critical that an organization come to consensus regarding the metrics to be universally utilized. The term "metrics" in this context refers to the specific measures to be employed and the units for those measures.

Various Views

In Section 1, I mentioned that practitioners in various disciplines within an organization might well perceive and define risk in very different ways. For example, employees in the area of health and safety usually interpret risks as threats to be identified and eradicated. People in the area of finance, however, are likely to see risk as an opportunity to be captured—low risk/low reward, higher risk/greater return.

So, it should not be surprising that in order to express their view of risk, members of various disciplines use divergent measures and formats for presenting those measures. Just a few of the measures and formats are shown in Figure A.1.

Lawyers, for example, are likely to express their perceptions of legal risks in the form of text. Managers of parts of a project might utilize a risk register to record, track, and present risk information. Engineers and other technically oriented folks might employ vehicles such as cumulative frequency plots to deliver the risk message. Others might utilize the "traffic-light" (red/yellow/green) approach, the Boston Square (*not* recommended), or any number of other measures and formats.

Getting It Together—Risk Monetization

Because a political risk can torpedo a project just as effectively as a technical, commercial, financial, or logistical (you get the picture!) risk, it is essential that the value of a project reflect the impact—both positive and negative—of the holistic set of threats and opportunities. This, in turn, means that no matter what metrics or formats were used to initially express the individual risks, all risks need to be translated into a single risk metric or a set of agreed-upon metrics.

I will consider the example of risks in a project. Typically, each project has an economic model. This usually is some giant unwieldy spreadsheet that takes in raw data such as time, CAPEX (capital expenditures), OPEX (operating expenditures), price for a product, efficiency, and other run-of-the-mill input parameters. Outputs from the model usually are just as mundane—NPV (net present value), IRR (internal rate of return), schedules (typically in the form of a Gantt chart), and other such indices and plots.

So, now the task is to create a Monte Carlo model that will allow us to feed the impacts of the risks to the spreadsheet economic model (SEM) so that

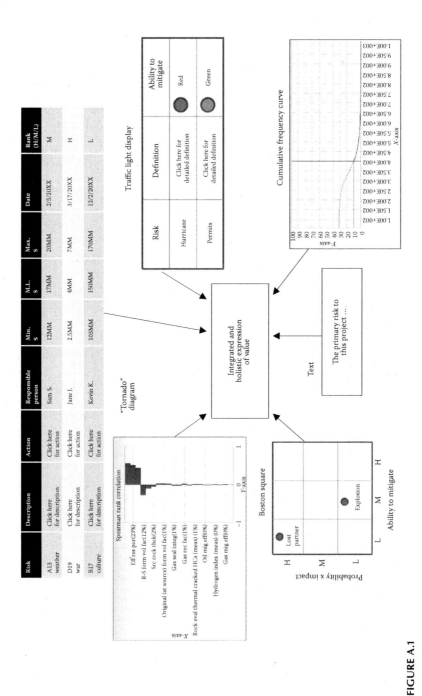

FIGURE A.1

Example of the expressions of risk that must be integrated to get a holistic view of perceived value for a project.

the output metrics are appropriately impacted. For those readers who are not starting out with a predefined economic model and who have to derive and build the model, or for those who need a layman's description of Monte Carlo analysis, please see my book *Risk Assessment and Decision Making in Business and Industry: A Practical Guide*, 2nd Edition in which I describe in detail the process of model building. For examples of real-life models representing myriad business, legal, and other situations, please see my book *Risk Modeling for Determining Value and Decision Making*. References for both books are found at the end of this appendix.

Let's imagine that someone has identified a political risk for a project. The risk might be stated like this: "Government red tape might delay the issuance of critical permits. If the permits are not issued on schedule, significant CAPEX will need to be spent to implement back-up plans."

Each risk should have an assigned risk owner (see the sections "Mitigation- or Caption-Action Owner" and "Mitigation- or Capture-Action Implementer" of this book and my book *Modern Corporate Risk Management: A Blueprint for Positive Change and Effectiveness* referenced at the end of this appendix for details on implementing risk processes in existing organizations). The risk owner should be a person who knows most about the risk. The political risk might have been first expressed using, say, the traffic-light format. A red light might have been assigned to this political risk.

Overall, the idea is to translate that "red dot" into something that can be used as input to our SEM. In a conversation with the risk owner, we might establish ranges (minimum peak and maximum values) for pertinent parameters so that, ultimately, the risk can be used to impact the value of the project through the SEM. So, in our conversation with the risk owner, we might first ask:

What is the range of probability that this political problem might happen?

A facilitated conversation would result in a minimum, peak, and maximum percent chance of the problem materializing. Next, we would ascertain the range associated with the query:

If the political problem does occur, what range of days (or weeks or whatever "time chunks") of delay might we experience?

The answer to this question should result in a minimum, peak, and maximum number of days of delay. Next, we would generate a range for the answer to the question:

How much money—per day of delay—might we spend?

A minimum, peak, and maximum value of dollars per day should result.

From the three ranges, three distributions can be built (see *Risk Assessment and Decision Making in Business and Industry: A Practical Guide*, 2nd Edition to see how distributions might be built). These distributions would be used as input to the Monte Carlo equation:

Political Risk CAPEX = (Range of Delay) × (Range of $/Day)

Using the probability distribution to decide whether or not the political problem should be applied on any given iteration of the Monte Carlo process, this range of additional CAPEX can then be appropriately applied (usually distributed over a time series) to the SEM. Such application will appropriately impact SEM-output measures of value such as NPV, IRR, and the like.

Another risk might relate to logistics. Such a risk might be stated thus:

> If freezing weather arrives more than one week earlier than expected, frozen waterways could delay delivery of critical materials, thus impacting the project schedule.

In a similar fashion to the process used to establish the CAPEX impact of a political problem, a facilitated conversation would be had with the appropriate risk owner(s). In that conversation, we would establish minimum, peak, and maximum values for ranges associated with the answers to

What is the probability that the weather problem will materialize?

What is the range of delay—expressed in days—if we experience the problem?

Again, in the Monte Carlo process, the probability range would be used to determine whether or not a delay would be applied on any given iteration. Such a delay would push forward in time expenditures, production, and all other time-dependent parameters. Output parameters such as NPV, IRR, and so on would be appropriately impacted.

In this example, we have chosen (because of the construction of the SEM) to make our consensus metrics the output parameters of the SEM (NPV, etc.)—typically expressed as cumulative frequency curves. Any parameter into which risks can be translated can be used. We refer to this entire process of translation of risks into impact on value as "risk monetization" even though the initial translation of a risk—for example, into probability or days or other things—might not be a direct translation into money. However, it is typical that some expression of value (money) is the ultimate expression.

In the next section, I will address how risks can individually and in confluence impact the perceived value of a project, how risks can be displayed, and how the entire risk-assessment process can be the cornerstone of the risk management decision-making process.

Section 5: Impact of Risks on Perceived Project Value

With the writing of this fifth section, I consider that our journey down the risk road is about half over. So, this seems a reasonable time to review the road we have already traversed.

Brief Review

In Section 1, I defined critical terms, outlined how risks can be identified, described, and managed, and introduced the fundamental drivers for observed human behaviors related to risk assessment and management. I concluded that section by proposing steps that might be taken to modify behaviors. In Section 2, I delineated in detail many of the behaviors that, in a company, impede the introduction and implementation of a comprehensive risk-assessment/management process. Section 3 centered on communication and the importance of establishing a common risk-related set of terms and definitions. Tips were given regarding how to establish such a glossary and the importance of doing so. The theme of Section 4 was metrics. I reviewed the disparate views of risk taken by various parts of a corporation, the different formats that can be used to express risk, and the fundamentals of the risk-monetization process.

In this fifth section, I attempt to elucidate how risks can individually and in confluence impact the perceived value of a project, how risks can be displayed, and will introduce the concept of how the entire risk-assessment process can be the cornerstone of the risk management decision-making process. The risk-management process will be the focus of Section 6.

In Section 4, I described how we might translate a political risk, for example, into metrics that could be used as input to an SEM. For the sake of consistency, I will repeat that translation here. The threat was the following:

Government red tape might delay the issuance of critical permits. If the permits are not issued on schedule, significant CAPEX will need to be spent to implement back-up plans.

Facilitated conversations with the risk owner would ensue to establish ranges for

- Probability that the threat will materialize
- Delay (in days or weeks or whatever "time chunks" are pertinent to the SEM)
- Cost per time period

Risk Monetization

On each iteration of a Monte Carlo process, a probability value would be drawn from the probability range and tested against an iteration-specific random number (between 0 and 1, or, between 0 and 100). If the random number is equal to or less than the probability value drawn from the threat-related probability distribution, then the threat is considered to have "happened" and a political-threat cost value would be calculated using the following simple equation:

Political Risk CAPEX = (Value from Delay Range) × (Value from Cost/
Time Period Range)

We would apply the impact of each risk to the SEM (see Section 4 for an example of a schedule threat and how it can be "monetized" to impact the SEM). In the Monte Carlo model, we can apply each of these risks one at a time. We are assuming here that the risks are independent—that is—not correlated. Correlated risks can be handled, but that process is too lengthy to be addressed in this short section. In this book, see the "Correlations between Threats and Opportunities" section of Chapter 2, the "Correlated Risks" section of Chapter 8, and my book *Risk Assessment and Decision Making in Business and Industry: A Practical Guide*, 2nd Edition for direction with regard to correlated risks.

In Figure A.2, we can see the result of applying the impact of each risk. The SEM-deterministic (i.e., single-valued) NPV is labeled "Base Case" and has, conveniently for this example, a value of $20 MM (20 million dollars).

Each risk is applied, one at a time. See the section "Viewing and Interpreting the Results of the Risk-Monetization Process" of Chapter 8 of this book for a full explanation. The data shown in Table A.1 is simply the impact—in NPV—of each threat and opportunity shown in Figure A.2. The value in the table is the difference between the base-case NPV and the probabilistically affected mean impact of the risk.

I have performed many of these types of risk-monetization processes, and the project team is almost always amazed at the value impact of the risks. A detailed description of this monetization process—in story form and layman's language—is given in Chapters 6, 7, and 8 of this book.

NPV-impact values in Table A.1 clearly illustrate where money and effort should be spent. This type of analysis is invaluable to project teams in directing their work and expenditures. Ranking risks based on impact on value—rather than, for example, cost—is critical to project success.

Introduction to Risk Management

Well, we have now just about come to the end of our risk assessment phase of these sections. I always liken the RA/RM link to a person having to cross a yard in which there is a vicious dog. The risk assessment part might include queries such as

- When does the dog sleep?
- How fast is the dog?
- How fast am I?

And so on.

Once you have the answers to all of the critical risk assessment issues, the risk management part ensues. Now, you might brainstorm ways to manage the situation, that is, how to get across the yard knowing what you now know. Some risk management possibilities might be

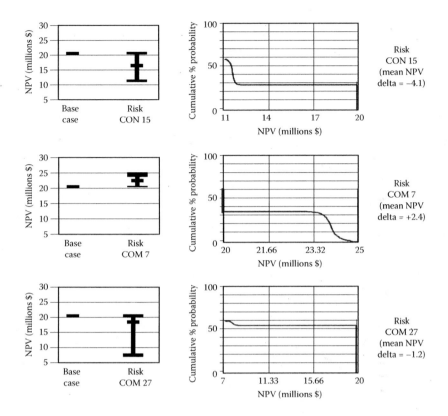

FIGURE A.2
The plots in the left are a variation on the stockbroker's "High/Low/Close-type" plot showing the base-case NPV and the range and mean (middle tick mark) probability-impacted NPV of three of the four risks utilized in the hotel-project example. The plots in the right are the actual cumulative frequency curves that are represented by the vertical bars in the leftmost plots. Delta NPV values are the difference between the deterministic base-case value ($20 Million) and the mean NPV of the risk's cumulative frequency range.

TABLE A.1

The Mean Probability-Impacted NPV Impacts—Relative to the Base-case Value of $20 Million—for the Two Threats and One Opportunity Considered in the Hotel-Project Model

Risk ID	NPV Impact (Millions $)
COM 7	+2.4
COM 27	−1.2
CON 15	−4.1

Note: Impact values for each risk were properly discounted and distributed in the time series for each risk, thus accounting for the time value of money.

- Wait until the dog falls asleep and take my chances.
- Drug the dog.
- Somehow distract the dog.
- Don stilts and saunter across the yard.

And other really good ideas.

Obviously, this is a boneheaded example, but it does clearly illustrate the difference between risk assessment and risk management. As I said above, we are now about done with the risk-assessment diatribes; in Section 6, I will begin to address the somewhat complex world of risk management.

Section 6: Risk Management and a Return to the Roots

Readers of the previous sections of this appendix know that in the first two sections I focused primarily on the necessity of changing behaviors and the attendant challenges. In Sections 3 through 5, I mainly delineated aspects of risk assessment. In this section, in order to address aspects of risk management, I will have to "return to my roots"—that is, I will revert a bit to the discussion of human behaviors and the complexities of organizations.

As you probably have noticed, I am a "splitter" rather than a "lumper" when it comes to considering risk. I make a distinction between risk assessment and risk management. I closed the previous section with a diatribe about needing to get across a yard in which there is a vicious dog. The story attempts to make the point that before we can manage something—including a risk—we must first have assessed the beast. The steps we might take in management depend almost entirely on a concise and accurate assessment.

The subject of risk management has been elucidated well in many prominent books. I direct readers to *A Guide to the Project Management Body of Knowledge*, 3rd Edition (PMBOK) which, with regard to risk management, contains the following sections:

- Risk Management Planning
- Risk Identification
- Qualitative Risk Analysis
- Quantitative Risk Analysis
- Risk Response Planning
- Risk Monitoring and Control

Another excellent guide to risk management is the book *Project Risk Analysis and Management Guide*. Many other books address the subject.

In these guides, the fundamentals of project risk management are described. In *Project Risk Analysis and Management Guide*, the risk management steps are presented as Initiate, Identify, Assess, Plan Responses, and Implement Responses. Presentations of risk management in other texts are, essentially, variations on this theme. Given that risk management processes are so eloquently described in many texts, it would be folly for me to attempt to recount even the essence of risk management in these short newsletter articles. So, then, why is this not the last sentence?

It's Not What to Do—It's How to Get It Done!

The texts listed above and other books on the subject do an excellent job of relating to the reader the steps associated with risk assessment and, mainly, risk management. When, lo those many years ago, I began to practice the assessment and management of risk, it became blatantly obvious right off that the main challenges were not procedural or technical, but were behavioral, cultural, and organizational. I noted that there existed plenty of books that addressed the processes/techniques of assessment and management of risk, but none that focused mainly on challenges and advice relating to getting people to *want* to implement the recommended risk processes. My book *Modern Corporate Risk Management: A Blueprint for Positive Change and Effectiveness* tries to address the issues related to implementation of the risk methods. So, in these remaining sections, I will focus on how to get people and organizations to take up risk management.

It is important that those who will be enacting the risk process believe that they will benefit from adherence to and implementation of the process. The existing incentive (reward) system might need to be modified so that people realize benefit from compliance.

Incentive systems usually work counter to risk-process implementation. For example, the cadre of bodies responsible for executing the initial steps in a project's lifecycle is under time and budget pressure. If a risk-identification workshop is held early in the project, the group of individuals responsible for the project at that stage of its lifecycle will be facing a litany of threat-mitigation and opportunity-capture actions.

Using threats as an example, a threat-mitigation action is a step that can be taken "now" in an attempt to assure that the threat will not materialize "down the road." Remember, the people responsible for the project at this early stage are under considerable stress to get their jobs done on time and on or under budget. Enacting mitigation steps will take time and money. Most of the threats that would be mitigated might be those that would impact temporally later stages of the project—stages for which the current group of individuals will not be responsible. If you were the manager of this initial group of individuals, what words could come out of your mouth that would convince the group that they should spend time and money to mitigate threats that might or might not happen and which certainly will not happen "on their watch"? "Take one for the team" is not likely to be a persuasive argument!

The bottom line is that if the incentive system is not modified to align with the risk process, then there is little hope that project personnel will act in a manner that is counter to the existing reward structure. Understanding the existing reward system and how the risk process will interact with that system is essential.

A Few More Hurdles and Proposed Solutions

In previous sections of this appendix, I have reflected on risk-management-implementation hurdles such as the reward system and inconsistent language. In this section, I will add to the list of implementation impediments one more salient stumbling block—that of the inability to implement. In subsequent sections, I will address and propose solutions for other major problems.

Although people in most corporations are aware of the things they should do, know how to do those things, and actually wish to do those things, they sometimes fail to execute.

Cultural, organizational, procedural, and other trammeling impediments provoke implementation paralysis. This inability to implement stems from many sources, just a few of which are time pressure, cost pressure, internal misunderstandings, and proven value and lack of support. I will briefly address the latter two issues, but a much more complete description of these maladies is available in the book *Modern Corporate Risk Management: A Blueprint for Positive Change and Effectiveness.*

I have in previous sections described the internal misunderstandings that can emanate from failure to establish a consistent and universal glossary of terms (i.e., language) related to risk assessment and management. I have proposed the "Standard German" solution for this potential source of misunderstanding.

Yet another barrier to implementation can come from misaligned or misunderstandings about expectations. A proponent of a risk methodology might poorly communicate, consciously or unconsciously, the time, cost, and organizational upheaval inherent in risk-process implementation. Project managers can be surprised by the real-world price tag associated with the effort, and typically resistance is the result. To alleviate this situation, clear, concise, "un-embellished," and episodic (but often!) communication with stakeholders is the successful path. Doing one's homework with respect to the needs and expectations of the organization can help prevent disappointments and misunderstandings.

The inability of an organization to implement risk practices can also result from lack of support from upper management. A dearth of backing can, in turn, be the result of management's reticence to support a process that does not have a proven track record of success. That is, the proposed risk process does not have proven value.

Nothing sells like success, and to that end, the proponent of a risk-assessment/management process should take the time, as practicality allows, to gather relevant testimonials-about and examples-of successful implementation. Prior to any effort to assail the lower echelons of the organization with the benefits of adopting a risk approach, such testimonials and examples should be used to obtain real backing for the project from those who will have to fund and staff the effort.

So, I'll close this section by reiterating that success in the risk management business is not so much about knowing what to do—lots of books can tell you that. The wellspring of success is the ability to get people, and the organizations that those people compose, to actually view the uptake of the risk process as beneficial to them and to the organization as a whole. In the next section, I will address more solutions to implementation problems.

Section 7: More Impediments to Risk Management/ Monetization Implementation and Attendant Solutions

In Section 6 of this appendix, I stated: "When I began to practice the assessment and management of risk, it became blatantly obvious right off that the main challenges were not procedural or technical, but were behavioral, cultural, and organizational. My book *Modern Corporate Risk Management: A Blueprint for Positive Change and Effectiveness* tries to address the issues related to implementation of the risk methods. So, in the remaining sections of this appendix, I will focus on how to get people and organizations to take up risk management."

In Section 6, I indicated that we typically get rewarded for successfully launching projects, but not necessarily for launching successful projects. I noted that in previous sections I have reflected on risk-management-implementation hurdles such as the reward system and inconsistent language and would in Section 6 focus on another implementation impediment—that of the inability to implement. In this seventh section, I will add to this list the challenges related to a lack of risk monetization.

Although the term "risk monetization" might suggest that a risk is converted directly into money, that is not necessarily the case. Corporate decisions are largely based on consideration of money in one way or another, so, in the end, risks should ultimately impact the perceived value of the project. However, along the way, the risk might initially be converted into a "nonmoney" parameter such as time delay, reputation impact, or another metric that is not measured directly in monetary units such as dollars.

In the fourth section of this appendix, I detailed just what the monetization of risk is and how to go about monetizing nontraditional risks such as political problems. To be brought up to speed with respect to the techniques involved in translating a risk into an impact on the perceived value of a project, I refer the reader to the fourth section

Risk Monetization—Great to Do but Difficult to Implement: A Perceived Focus on Threats

It should be obvious that if the perceived value of a project can be significantly influenced by the impact of threats and opportunities (i.e., risks) and that if the decision regarding whether or not to proceed with a project depends largely on the value the project will bring to the company, then any decision maker would desire or require that the project value on which a decision will be made should reflect the impact of all pertinent risks. That's what you'd think, but you'd be wrong.

In the technical vernacular, there are "loads or reasons" why the impacts of risks are not reflected in the perceived value of a project. The first might be the perception that a focus on risk is a focus on the "downside"—that is, the emphasis of threats.

It is mainly true that when a monetization process is completed, the perceived value of the project has been diminished relative to the before the impact of risks value. It has been my experience that when risk-identification processes are finished, the list of risks includes mainly threats and relatively few opportunities. This phenomenon is not the result of a focus on threats. Risks related to a project are relative to the base case. That is, when the risk-identification process begins, the project manager or commercial analyst presents to the identification-process participants exactly the project that is being assessed. That project description typically includes most of the attendant opportunities—even most of the low-probability events—and few threats. Therefore, given that the base-case description for the project has already taken account of almost all of the "upside," it is no wonder that a risk-identification process can uncover mainly threats that should impact the base case and project value.

The remedy for this pervasive malady is to attempt to influence projects as early in project life as possible. Moderate- and low-probability opportunities might be kept out of the base case and the inclusion of relatively high-probability threats should be included. In this way, a more realistic base case is first put forward and surprises down the road are minimized. Such a project base case will, in the risk-identification process, yield a more balanced list of threats and opportunities. It is the job of the RPP—the person on the project staff who is charged with shepherding every aspect of the risk process—to attempt to ensure that a realistic initial base case is created. Assignment of a trained RPP is critical to project success. More about the role of the RPP can be found in the first section of this appendix and in any of the books listed at the end of this appendix.

Risk Monetization—Great to Do but Difficult to Implement: Not a Line Item

If a project is "killed" somewhere along the attendant timeline, it is mainly due to risks—primarily threats—that do not appear as "line items" in any project economic model or analysis. For example, it is common for any project-analysis spreadsheet to include rows or cells for such things as production, capital expenses (CAPEX), operating expenses (OPEX), price, utilization, and the like. It is not typical, however, to find a row or cell specifically dedicated to "politics" or "organizational capability" or any other "soft" risks.

I can tell you from experience that it is these "soft" risks that represent the greatest threats to any project. A governor of a state, for example, might, part way through a project, decide to modify the tax structure precipitating a salient economic downturn in project value. Perceived project value might also suffer diminution at the hands of environmental risks. Part way along the project timeline, it could be discovered that critical permits will not be issued by a governing body unless/until important environmental concerns are successfully addressed. Although organizational capability might not exist as a line item in a project economic spreadsheet, it is not uncommon to discover that critical departments and disciplines within a company lack sufficient synergy and communication to successfully execute the project.

It is the responsibility of the RPP to ensure that such "soft" risks are, to the extent possible and practical, identified and succinctly described in the risk-identification process. This is the only reasonable remedy for the "soft risk" problem. Cogent description of risk response plans and succinct monetization of such risks should be emphasized by the RPP.

Risk Monetization—Great to Do but Difficult to Implement: No "Pull"

It has already been pointed out in this section that the risk-monetization process is largely viewed as an exercise that will lessen the perceived value of a project. As I have lamented in previous sections, the prevailing incentive system rewards project leaders for successfully launching a project but not necessarily for launching a successful project. In such an environment, it should come as no surprise that there might be little to no "pull" from upper management to implement a risk-monetization process.

If management "above" the project level expects to make decisions based on the best possible estimate of project value, then one remedy to this pervasive problem is to influence decision makers such that they expect and demand that risk impacts have been taken into account. Another solution to the "lack of pull" challenge is to convince project leaders that it is in their best interest to embrace the risk-monetization process.

One of the outputs of a risk-monetization effort is a table that lists the impact on value of each risk. This is an indicator of the "worth" of each risk and, in turn, indicates what might be spent to mitigate a threat or capture an opportunity (you don't want to spend a dollar to offset a dime). Project leaders are under tremendous pressure to stay within budget. Risk monetization can be touted as one method that will help cash-strapped project leaders to address the risks facing the project in the most economic manner.

In the eighth section of this appendix, I will address other impediments to implementation of a risk-management process in any organization. The following ninth section and the final section will focus on aspects of risk processes that will help any project team member better utilize risk-based information in the successful implementation of a project.

Section 8: Edicts and the Letter and Spirit of Risk-Process Implementation

In closing the seventh section of this appendix, I addressed the critical aspect of gaining support from upper management for implementation of any risk-assessment/management process. Although such endorsement is essential, this is truly a case of "beware of what you ask for."

"Thou Shalt ..."

Much of the material on which I draw to write these articles emanates from personal experience. The alignment of upper-management support is no exception.

Years ago, I served as a research scientist at a large technology center. Toward the later stages of my tenure at that institution, I was primarily focused on the development of risk-based technologies and their implementation in the corporation. By happenstance, I was friendly with the president of the company who, as luck would have it, also took an interest in things risky. He arranged for me to have two audiences with the CEO.

At the second of those presentations to the highest of management, the CEO said (and I paraphrase, here): "OK, you convinced me that we should implement risk-based processes in the company. So, why don't I just issue an edict indicating we will start doing so?" I immediately tumbled to the realization that I had got what I wanted, and now understood that it was not what I wanted.

I was fully cognizant of the challenges and barriers associated with injecting risk-based processes into the everyday workstream of most disciplines. I knew that if the CEO issued an edict such as "Thou shalt do risk assessment and risk management," we would be in big trouble.

Why trouble? Well, first, it was a big company. Just about 100% of the risk expertise and resource on which the company could draw was standing in front of the CEO—that is, me. If, suddenly, all departments in the organization felt compelled to do something related to risk, how would they know what to do? How would they get the critical information, training, and support it surely would take to bring risk-based processes into their part of the business?

One answer to this mainly rhetorical question is that they look outside the company for consultants who claim to know something about risk. Unlike chemistry, physics, English literature, accounting, and other courses of study at most universities, there is—at the time of this writing—no "risk" major that is recognized and accredited. So, anyone who wishes to call themselves a "risk expert" can do so—and they do! It is absolutely true that any company is better off not employing risk-based processes than having the mosaic of disjoint and questionable-quality risk processes that will result from opening the corporate doors to anyone (and some are real snake-oil salesmen!) who can make a buck in trying to "help" your company.

At that presentation to the CEO, I was quite embarrassed to have to back-peddle and do what I could to convince him that a "Thou shalt" edict was absolutely not in the best interest of the company. Following the presentation, I was forced to ask myself: "OK, if I didn't want the CEO to issue an edict, just what did I really want?"

Turns out that implementation of risk-based processes in any organization is a delicate dance and a chicken-and-egg situation. As the RPP, you certainly need upper-management support for the roll-out of risk applications. However, before asking for such high-level support, you have to have in place the resources, materials, and expertise that will be necessary when you, hopefully, get the "go ahead" from the upper echelon. Without such support to begin with, however, how do you garner the money, resources, and time it takes to prepare for the onslaught that will surely come when upper management gets behind the idea?

The only resolution to this dilemma is to make sure that when attempting to convince upper management that such risk processes are necessary, you emphasize that much time, money, and resources will be necessary to build an organization and technologies that will be necessary to handle the rigors and demands of implementation. I learned this lesson the hard way. I hope from these few paragraphs, I have saved you from attending the school of hard knocks! Examples of such implementations can be seen in the book *Risk Modeling for Determining Value and Decision Making* and much more about the problems and related solutions related to process implementation can be found in the book *Modern Corporate Risk Management—A Blueprint for Positive Change and Effectiveness.*

The Letter but Not the Spirit

OK, suppose you have successfully navigated the minefield of obtaining management support, building your support group, and developing all of the training and implementation materials that will be needed for a successful roll-out of the risk-based processes (whew!). One other way this thing can backfire on you is that those in the company on which the risk processes are foisted comply with the "letter of the law" but not with the spirit.

For example, I have often observed that minimum requirements of a risk-based process are (at least) the following:

1. Hold a risk-identification event (at which threats and opportunities are identified and roughly ranked).
2. Generate a risk register that lists all of the threats and opportunities.
3. Create and record in the risk register threat-mitigation and opportunity-capture plans along with due dates, risk owners (names), etc.
4. Assign an RPP who will "ride herd" on the entire process (make sure those responsible for taking actions actually do so).

Although these are necessary steps, unless such steps are combined with, at least, some aspect of risk monetization (see previous articles in this series for much more on the risk-monetization process), you can actually be doing the organization a disservice by promoting the steps listed above.

If the "letter of the law" is followed but not also the spirit, implementation of risk-based processes can be a significant waste of time, money, and resources. I have often observed parts of organizations that, for example, implement the four items listed above. Long lists of risks are created (the risk register) and for each risk, mitigation or capture actions are defined and those people responsible for executing the mitigation/capture plans are identified. This often results in many folks expending much effort in trying to carry out their tasks. This might seem, again, like what you would want, would it not?

Without integrating, just for example, even a fledgling risk-monetization process (the spirit of the law) with the risk-register process, many of the risks that are addressed are those that are of minor impact that likely should have been ignored. Without someone, such as a skilled risk-monetization expert, coming along to determine the actual value impact of each risk in the register, it is not necessarily possible to determine just which risks should be addressed and which risks should have no effort put into related mitigation/capture plans.

Having everyone running around attempting to "fix" every risk is, unquestionably, a significant waste of corporate resources. The lesson to be learned here is that it is not enough to identify risks and associated responsible parties and mitigation/capture plans. If such steps are not linked with value-impacting processes and a high-level ranking of risks based on value, implementation of risk processes can actually be a step backward rather than a forward stride. The technologies utilized to implement such processes are detailed in the book *Risk Assessment and Decision Making in Business and Industry: A Practical Guide*, 2nd Edition.

In the ninth section of this appendix, I will focus more on how to practically implement remedies to many of the problems I have outlined in previous articles. Ah, finally a focus on the solutions rather than the problems!

Section 9: The Road to Success

Well, in almost all of the past sections of this series of 10, I have focused primarily on the problems, hurdles, and challenges associated with implementing a risk-based process in any organization. In this ninth section, I will stress five steps to success that I deem to be essential.

1—Make Sure They Get It!

In Section 8, I stressed that the support of upper management for the risk-based effort is absolutely critical. I will add here, that such support has to be genuine. I have many times experienced upper-management "support" with said management believing that the changes necessitated by embracing a risk-based business process would impact those personnel "down the ladder" and would not make meaningful differences in their everyday responsibilities. This very-wrong interpretation typically accounts for their quick agreement to support the effort ("Great, as long as it is for everyone else and it does not impact *me*!").

I experienced this situation firsthand in Egypt years ago (one of many such experiences). The "Big Manager" on location had called us to Egypt to bring risk-based processes to his exploration/production effort and made, while we were present, great pontifications to the gathered employees regarding just how important it was to embrace the new risk-based process. Of course, probabilistic analysis results, usually, in results expressed as ranges and probabilities (a typical cumulative frequency curve, for example). Later, when speaking with one of the "Big Manager's" direct reports, that person indicated that when bringing information to the "Big Manager," you had better not show up with a range of answers—he wanted *THE* number! Just another example of saying one thing but meaning another. This subject is expanded upon greatly in the book *Risk Assessment and Decision Making in Business and Industry: A Practical Guide*, 2nd Edition.

The lesson here is that management has to be prepared regarding the impact that a switch to risk-based processes will actually have on them and the way they do business. This leads me to the education aspect of this section.

2—Education at Three Levels (at Least)

This educational aspect is part of the "chicken-and-egg" nature of implementing risk-based processes in any organization. Preparing classes/

seminars for multiple levels of corporate personnel can be time consuming. Your management—those paying for the risk-based-process implementation—have to be prepared to spend considerable resources in preparation. Generating targeted classes/seminars are part of that preliminary preparedness.

First, you have to do much homework regarding the businesses to be addressed. Presenting generic examples in class regarding risk-process implementation will not "cut it." Risk-based examples absolutely need to be taken from the business being addressed and couched in their terminology. This can mean quite a bit of investigation on your part so that a credible set of examples can be created.

In addition, courses should be prepared for three separate audiences. These are

- Top Business Management. For these few folks, a relatively short and to-the-point less than one-day presentation should be created. It should include the following:
    1. What will be the benefits to their business from implementing the risk-based approach.
    2. What will their personnel be using as input to the risk-based system and what the output means.
    3. What will this effort cost in terms of time, human resources, and money.

- Business Unit Management. These are the individuals who will deal directly with the output from the risk-based system, and, will have to make smart business decisions in the new risk-based world. These folks will also be directly responsible for those personnel who are actually generating, running, and producing results from risk-based processes. A course for these managers should include the following:
    1. What will be the benefits to their business from implementing the risk-based approach.
    2. What will their personnel be using as input to the risk-based system and what the output means.
    3. What will this effort cost in terms of time, human resources, and money.
    4. What the output from the risk-based processes really means and how they can use such output to advantage in their business—this is very important and should be illustrated with real examples.
    5. How to present this new output and information to their managers and partners.

- Personnel who will actually run/use the risk-based applications. These are the individuals who will deal directly with the output from the risk-based system, and, will have to make smart business decisions in the new risk-based world. These folks will also be directly responsible for those personnel who are actually generating, running, and producing results from risk-based processes. A course for these managers should include the following:

  1. What will be the benefits to their business from implementing the risk-based approach.

  2. What will this effort cost in terms of time, human resources, and money.

  3. Exactly what information and type of information (ranges, for example) are required.

  4. Exactly what happens to those data when processed by the risk-based systems (remove the "black-box" effect).

  5. Extensive hands-on training with each risk-based system using examples and data that relate to their jobs.

  6. What the output from the risk-based processes really means and how they can use such output to advantage in their business—this is very important and should be illustrated with real examples.

  7. How to present this new output and information to their managers.

3—Survey the Risk Landscape

As mentioned in previous sections, the risk-based process being discussed in these appendix sections is one that is holistic. That is, it encompasses and integrates risks from almost all pertinent disciplines such as technical, legal, engineering, commercial, financial, construction, security, health and safety, environmental, logistic, and many others. There is no question that within an existing organization there exist several organizations that believe they already make use of a risk-based system and that their system is all-encompassing.

For example, it is not unusual to find an audit group or a health and safety group that use risk-based processes. These processes can be quite comprehensive within their specific area. Personnel within these specific disciplines typically cannot imagine a more sophisticated or holistic approach. They are correct, usually, if the scope of risk is limited to their area, but your charge is to integrate the data/process used by their discipline with risk-based processes and data from all other disciplines. This, they (the Health and Safety folks, for example) will not understand. Managers and personnel within these disciplines can be significant sources of resistance to your more comprehensive effort because they believe "they have it covered" and very much feel threatened by your much more broad-spectrum approach. Examples of how differently disparate groups can view the subject of risk are given in the section "Many Views of Risk" in Chapter 1 of this book and in the book *Risk Modeling for Determining Value and Decision Making.*

So, before attempting to run roughshod over the already-existing risk processes, make sure that you make considerable effort to learn about those existing processes and to take plenty of time to meet and communicate with these disciplines. Personnel using existing, but colloquial, risk-based systems need to know that they are NOT threatened by the new comprehensive process and, in fact, the input from their existing system will be a critical part of the new effort. This, of course, has to be true. So, many comprehensive and holistic risk-based processes are killed before they get established due to internal sabotage emanating from already-existing entities that feel threatened.

4—The Common Language Thing Again

I have addressed this aspect of risk-process implementation in several of the previous eight sections of this appendix. It keeps coming up because it is of such salient importance.

Miscommunication is the bane of any universal effort, and this is especially true for risk-based initiatives. In previous sections, I have described in detail the setbacks that can be expected when and if various parts of the organization utilize colloquial verbiage with regard to risk. Typically, these cul-de-sacs communicate about risk only within their discipline and have little opportunity to attempt to communicate across disciplines. For example, the risk folks in the audit function might never sit down with individuals from law to discuss how their risk perceptions and terms and definitions compare. When a comprehensive risk-based approach is implemented, suddenly the definitions of terms matters. Without some annealing effort, miscommunication is assured.

I have related in an earlier section that it is folly to believe that you will change, for example, the way health and safety folks view and talk about risk. You should not waste your time in an attempt to cause them to modify their ways. I have suggested the "Standard German" approach to this problem. That is, don't attempt to change the local dialects, but create a set of agreed-to terms and definitions that, when used for universal communiqués, will be understood by all—even though none of the disciplines might adhere to those terms and definitions within their particular disciplines (nobody actually speaks "Standard German," but all people speaking all dialects can read the same newspaper articles written in Standard German). In this way, various independent entities can clearly communicate across boundaries. This works! Do not forget to address the communication issue!

5—The Reward System

I have saved the most important aspect for last. As mentioned briefly in Sections 1 and 6 of this appendix and as delineated at length in my latest book, *Modern Corporate Risk Management—A Blueprint for Positive Change and Effectiveness*, regarding project teams, it is mainly true that we get rewarded for successfully launching a project, but not necessarily for launching a successful project.

This concept, and the inadvertent implications of most incentive systems, has so many facets and aspects that it's just a bit silly to try to adequately recount it in this article. The reader is referred to the aforementioned text for more detail. However, a few of the major "fixes" for the dark-side symptoms of the reward system can be described here in at least a cursory manner.

It is not uncommon to run across incentive systems based on "saving." That is, rewards are distributed for "saving" money or "saving" time, and so on. In a project system that is stage-gate driven, or, in a system in which a project, during its lifetime, is passed from one corporate entity to the next (e.g., from the negotiators to the commercial arm to the design department to the engineering department to the construction department . . .), incentives based on "saving" typically result in poor project results. If corporate-entity #1 is rewarded for saving money, then it might be true that when corporate-entity #2 inherits the project, it will receive a less than optimally prepared project. In the end, the operators of this project typically pay a high price for attempting to operate and maintain a facility, for example, built on "the cheap."

The lesson to be learned here is to design the reward system with the end-game in mind. If the corporation actually wishes that the project process result in a well-conceived, adequately designed, and decently built entity, then incentive systems should be put in place that reward the quality of the end result and that do not reward intermediate constructs. This, of course, might mean delaying the reward for those involved early on requires a good "corporate memory" (who did what, when), and can mean that groups that did an exemplary job in the early stages can get robbed of their reward by subsequent groups that "dropped the ball." One way to attempt to ensure that the quality of the project is at least adequate at the handoff from one group to another is to employ a qualified but disinterested party to pass judgment on the adequacy of the work being passed along.

Just one more of a long string of reward system foibles relates to the practice of employees tying their sense of job security to the success of the project. In Section 2 of this appendix, I relate a story about assembly-line workers, but I actually prefer the "ugly baby" scenario. If you ask the grandparent if the baby is cute, what response do you suppose you will get (even if the child is a bit frightening!)? So, it can be with project team members.

I have had the experience of traveling to a foreign country to perform a risk assessment on a project. The team members have moved to the country, moved their families, put their children in schools there, have purchased real estate, and so on. I do not intimate here that these folks are dishonest, however, there is no question that their view and expressed opinions about the project are tainted by the fact that they tie, rightly or wrongly, their job security to the success of the project. When interviewing these folks regarding critical data to be input to the risk analysis, it can be a supreme challenge to obtain objective information.

One "fix" for this syndrome is to disassociate most of the project personnel from the project's success. For example, rather than sending a construction engineer to the foreign country on a permanent basis, such talented individuals can be seconded to the project from a central pool of engineers. These folks will be in part reviewed by local in-country management, but their promotions and so on can be actually controlled by the central-engineering organization. Such personnel can be rewarded for "calling 'em like they see 'em" with regard to problems on any given project—although, not to the extent that people are rewarded for making-up problems!

Overcoming the challenges of the reward system can be daunting. However, just like curing a disease, the first step is to recognize and admit that you have a problem. The reward system is *always* a problem and the preceding paragraphs outline only a few of the suggested steps that might be taken to resolve just a few of the most critical aspects.

In the tenth and last section of this appendix, I will attempt to encapsulate and summarize the highlights from all of the preceding articles. Stay tuned for the last of the installations.

Section 10: A Synopsis of the Risk Assessment/Management Journey

This is the last in the series of ten sections on risk assessment and management (RA/M). Just like implementation of an RA/M process in any organization, this series of short diatribes has been a journey. Of course, it is impractical to attempt to encapsulate the details of each previous section in this last one, but it is entirely feasible to attempt to relate the highlights of the journey. Below I have organized the synopsis into sections on organizational/cultural issues, technical aspects, risk management, and recommendations for successful implementation of RA/M.

Just Some of the Organizational/Cultural Stuff

To the uninitiated, it might seem as though RA/M is all about measuring threats and opportunities (i.e., risks) and implementing processes to mitigate (threats), capture (opportunities), and manage risks. Although measurement and management are secondary tenets of any RA/M scheme, it is really all about changing behaviors—of organizations and of individuals.

With regard to implementing an RA/M process in any organization, the statement "We get rewarded for successfully launching a project and not necessarily for launching a successful project" captures and encapsulates both the essence of the challenge and the basis for remedy. Sections 1 and 2 primarily addressed these issues.

After describing a risk as "A pertinent event—threat or opportunity—that has a textual description" and further asserting that a risk has at least an associated probability and consequence, I suggested that "proving" the benefit of RA/M was most easily done when RA/M is applied to a portfolio of relatively short-term projects. It also was proposed that a project team member's feeling of personal success and security has to be separated from the success of the project if objective RA/M is to be realized. Proposed was the "assembly-line worker" analogy. This model envisions creation of a workforce composed of individuals who are interested in creating the best portfolio of projects rather than tying their sense of security to a specific project. Reorganizing the workforce into discipline-specific central units was recommended. Members of these units would be seconded to multiple projects as a means of breaking the tie between a worker's perceived sense of security and the success of a particular project.

In the third section of this appendix, it was stressed that communication is key. Myriad disciplines will view, discuss, and measure risk in colloquial ways. I suggested that it is folly to attempt to change the language employed by any discipline but, rather, it should be agreed that a common set of terms and definitions will be utilized by all disciplines when communicating with one another, or, "up" or "down" the chain of command (the "Standard German" model).

Just Some of the Technical Stuff

The crux of Sections 4 and 5 was the idea that the various disciplines that contribute to successful execution of a project (legal, security, health and safety, environmental, commercial, financial, logistics, political, engineering, and so on) will not only communicate uniquely about risk, but each will measure risk in a colloquial manner and will express risk in a fashion that best expresses their measurements and communication mode. Risk registers, traffic lights, tornado diagrams, cumulative frequency plots, PIG (Probability/Impact Grid) plots, colored-box matrix, text, and a host of other risk-presentation methods will rightfully be employed.

It is the job of the RPP to ensure that these various measurements and expressions of risk are integrated and the holistic impacts of these risks are reflected in the perceived value of the project. The risk-monetization process is the recommended method for integrating the impact of threats and opportunities and for reflecting that consequence in perceived project value.

The risk-monetization process is essential for sorting the threats and opportunities in the risk register so that only those risks are addressed that will enhance value.

Just Some of the Risk Management Stuff

Sections 6, 7, and 8 focused on risk management and I began by espousing that the impediments to risk management are primarily behavioral, cultural, and organizational. Recommended risk-identification workshops—held early in the project lifecycle—will identify risks that might occur "down the road" and "on the watch" of individuals and groups different than those who have control of the project at the time of risk identification. One of the main changes to the reward system centers on encouraging project team members—who are in control of the project when the "down the road" risks are identified—to take an active (and usually expensive) interest in implementing steps to mitigate threats and to capture opportunities. Many threats and opportunities typically are projected to occur in later segments of the project lifecycle when the contemporary project team members have long since relinquished control and responsibility. Cultural, organizational, and, mainly, reward system modifications are usually necessary.

It is pointed out in these sections that the RPP should be ready to deal with the erroneous criticism that risk-identification workshops focus on threats and give less emphasis to opportunity identification. This perception stems from the long litany of threats and relatively short list of opportunities that result from such workshops. The disparity in the magnitude of the number of threats and opportunities is a consequence of the tendency of project teams to have accounted for almost all opportunities in their "base case" leaving few opportunities to be identified. Threats, however, are mainly ignored when creating the base case and, therefore, the list of potential threats not accounted for can be relatively lengthy. It is the job of the RPP to attempt to ensure that, in the end, the impacts of all pertinent threats and opportunities are reflected in the perceived project value.

Yet another aspect of risk management is to attempt to ensure that "soft" risks are taken into account. A thing such as political risk is rarely a line item in a project economic evaluation. However, it can be things like political risk that have the greatest impact on the project.

One of the last major risk management points in Sections 6, 7, and 8 is that of how upper management perceives the RA/M process. It is communicated in these articles that it is essential to have upper-management "pull" for the RA/M process (and members of upper management have to understand RA/M and mean what they say!), but that too much "pull" can be disastrous. For example, if upper management were to suddenly become imbued with the spirit of RA/M and issued an edict to the organization that "We will now do RA/M," the issuance of such an edict can easily come before the RA/M support materials, educational classes, personnel, and so on are in place. Such a situation can result in the organization seeking "outside" RA/M assistance—most of which will not be of a caliber or of a philosophical posture that is satisfactory. Beware of what you ask for!

Just Some of the "Road to Success" Stuff

In Section 9, I attempted to consolidate many of the steps that might be taken to ensure a successful RA/M implementation. I addressed the upper management "getting it," communication, and reward system issues in detail—aspects of RA/M to which I had previously alluded. However, in Section 9, I described in some detail the areas of education and of surveying the risk landscape.

I suggest that timely education be offered at (at least) three levels. Separate educational efforts should be designed for Upper Management, for Business Unit Management, and for those personnel who will actually run/use the risk-based applications. Unless folks at these three levels truly understand RA/M and the impact it will have on them, on the organization, and on the fortunes of the company, it is likely that compliance with RA/M practices will lax.

A last point made was to be sure to survey the existing RA/M landscape of the organization before you attempt to foist upon them a new and different RA/M scheme. Tremendous resistance can be mustered from entities that, right or wrong, feel as though they already employ an RA/M process that adequately addresses their needs. Identifying such processes and "folding them into" your new RA/M proposal can be most advantageous.

So, I hope you have enjoyed reading these 10 sections as much as I have enjoyed writing them. All of the information conveyed, and much more, is described in detail in the books referenced below. Please feel free to contact me regarding any risk issues you might have, and thanks again for your time and patience.

References

A Guide to the Project Management Body of Knowledge, Project Management Institute, Inc, Newtown Square, Pennsylvania, 2008.

Bartlett, John, et al., *Project Risk Analysis and Management Guide,* APM Publishing, Buckinghamshire, U.K.

Koller, G. R., *Risk Modeling for Determining Value and Decision Making,* Chapman & Hall/CRC Press, Boca Raton, FL, 2000.

Koller, G. R., *Risk Assessment and Decision Making in Business and Industry: A Practical Guide,* 2nd Edition, Chapman & Hall/CRC Press, Boca Raton, FL, 2005.

Koller, G.R., *Modern Corporate Risk Management—A Blueprint for Positive Change and Effectiveness,* J. Ross Publishing, Fort Lauderdale, FL, 2007.

Index